GOD IS

BY

ALAN G. GREER

KCM PUBLISHING
A DIVISION OF KCM DIGITAL MEDIA, LLC

CREDITS

God Is by Alan G. Greer

ISBN 13: 978-1-939961-98-3
ISBN 10: 1-939961-98-X

First Edition

Publisher: Michael Fabiano
KCM Publishing
www.kcmpublishing.com

The KCM logo is a registered trademark of KCM Digital Media, LLC.

Praise for *God Is*

God Is, is a must read. Alan Greer takes us on a flowing historical journey from the Big Bang theory to the present. His scientific, military and legal backgrounds allow him to explain how the force of God is behind and the reason for, all that happens. He then warns us of what can happen if we don't remain the free willed students of God, who has taught us to share the attributes of love, patience, empathy, fidelity, caring, moral courage, trust, and the willingness to make the right, but difficult decisions. This is a call to all of us. BRAVO!

- Dr. Pedro J. (Joe) Greer Jr., MD, Associate Dean of the
Florida International University (FIU) Medical School
and the recipient of the U.S. Presidential Medal of Freedom

In *God Is*, Alan Greer makes the case — sweeping and impassioned, yet at the same time thorough and lawyerly — for the existence of a benevolent God. As a longtime atheist, I was challenged by his arguments and intrigued by his conclusions, which carry profound implications for the future of humanity.

- Dave Barry, NY Times bestselling author,
recipient of the Pulitzer Prize for Commentary (1988)

The author, deeper than I, has an eloquent touch for explaining what others might say cannot be explained. Alan Greer's *God Is,* is education at its best: Why none of us will find ourselves bored eternally in Heaven, a takedown of the Noah story, why we humans need to be challenged all our lives, God The Choreographer, God The Ballet Master, God The Loving Gardener, the imperatives of trust and patience. This book is a blessing in the pursuit of the greatest meanings of life — and death.

- David Lawrence Jr., chair of The Children's Movement of
Florida and the retired publisher of the Miami Herald

To ponder a God active in history and science one has to demonstrate a knowledge of history and science, and that Alan does in a succinct and systematic fashion. He then ventures into the ethereal realms of free will, soul and afterlife, and lovingly shares a personal theology that encourages a partnership with the Divine. In doing so, he encourages each of us to wrestle with God ourselves.

- Mitchell Chefitz, author of *The Seventh Telling: The Kabbalah of Moshe Katan*

Dedication

To my beloved
wife Patricia
whose deep faith
inspired so many
parts of this book.

Table of Contents

Foreword

For millennia, people have pondered the question of God, seeking to figure out the meaning and significance of life. We have the early sacred texts of many religions. Then there are the rabbinic sages of the first centuries in the Common Era, early Christian Church Fathers such as Augustine, Calvin, and Luther during the Reformation, and later Christian theologians as well as Jewish ones like Rabbi Abraham Joshua Heschel. Ordinary lay people have also thought about the problem of God. Though it is often said we live in a secular, post-religious world, religion is far from dead, and its influence is not always benign, as is obvious by the role religion has played in the escalating conflicts and divisions in the world, including America. But Alan G. Greer's new book offers his readers hope, as he strives to define God and God's role in the universe.

Humans cannot seem to get away from thinking and writing about God. Even self-proclaimed atheists like Christopher Hitchens spend a lot of time on the subject. Is there a God, or whatever one might call a supremely powerful force, something greater than ourselves, beyond our comprehension? Or is God something we have created, perhaps in our own image? If there is a God, what is God? A creator? A judge? A compassionate nurturer? Benevolent or malevolent, caring or indifferent, or all of the above? Is God beyond gender? If God is all-powerful, where do humans fit in and what is our role or do we even have one?

God Is is an affirmation, both passionate and rational, of the existence of God, and Alan Greer offers an argument for God's plan for the universe and humankind. Highly intelligent, widely read, a man with long experience and wisdom, Alan Greer constructs an argument that builds chapter by chapter. As befitting the distinguished trial lawyer that he is, Alan Greer puts God on trial. And guess what? God is exonerated. I am reminded of John Milton's epic *Paradise Lost* (1667), that canonical seventeenth-century English epic which opens with Milton boldly declaring his intention to "justify the ways of God to man." Alan Greer is not a poet but, like Milton, a layperson confronted by challenges particular to his time and inspired to make a case for God, for a "good" God.

I am intrigued by a lawyer defending God, against the prosecution's claims that God does not exist, and against the idea that God is absent or does not care about us. Greer's God is paradoxically both remote (God does not directly intervene) and present (as a "shaper" or a "silent voice"). God is not human yet has human qualities—intentions, love, patience. Alan Greer builds his case for God out of what he admits is "circumstantial evidence," concluding in a brilliant lawyerly move that "circumstantial evidence" may be even stronger than direct evidence.

In a world full of skepticism, and particularly in our present world where it does not seem a benevolent God is in control, this book is a ray of light, offering comfort even to readers who may not be convinced by everything Greer asserts. He is clearly a "person of faith." I am not sure I consider myself exactly a person of faith; I am a woman and a Jew who struggles with faith. I want to have it, at moments I think I do (though only God knows), but many things in this world make it hard for me to be sure there is a God in control of everything, that good triumphs, or to believe in progress. So I found it hard to accept some of Alan Greer's statements and assumptions, particularly in the earlier chapters of the book: that throughout time we have been advancing, that we have been moving towards "perfection," and that it is all part of God's plan. Greer's God "engineers" things, shapes his "garden" of the world, weeding out less good plants or species so that the universe can progress to perfection. All is for the good; even violence serves God's purposes. Such statements unsettle me. But the way in which Greer backs up his claims with arguments drawn from physics, astronomy, biology, anthropology, and history complicates these assertions and makes me think. His explanations of some of the stories of the Bible—particularly his reinterpretation of the Adam and Eve story of Genesis—are terrific. He makes the case that Adam and Eve's "fall" was actually God's "gift" to humans, thus implicitly but sharply rejecting Augustine's doctrine of "original sin," which has long influenced Christianity. But Greer's reinterpretation of the Adam and Eve story is also significant as it provides the foundation for his overarching argument that perfection and growth are ongoing processes, perhaps even beyond death.

Sometimes I found myself arguing with Alan Greer. How can he say God is so loving and has a plan for perfection when the world seems to be going to hell? Are humans really so special if they act so badly? Why does not God intervene if God is so good? As Greer argues for "intelligent design" in creating a "perfect chain of being," I kept thinking of Alexander Pope's *Essay on Man* (1734), which, hoping to "vindicate God's ways to Man," asserted "whatever is, is right!" Pope's optimism has always disturbed me, as it did his friend Voltaire, whose novel *Candide* satirized the idea that everything happens for the best, and that this is the best of all possible worlds.

To be disturbed. Maybe that is part of Alan Greer's purpose—to make us think, and argue, as if the resistant reader takes the place of the prosecutor, whose objections need to be overcome.

As I read on, though, I found myself increasingly fascinated. Greer carefully, thoughtfully fashions his answers, based on his knowledge of history and science. Particularly moving is his account of how in the catastrophic devastating "extinction event" at the end of the Cretaceous Age," remnants of life" survived that eventually would result in the evolution of *homo sapiens*. His point that something good survives, even comes from catastrophe, gives the reader hope--something we sorely need in our dark times.

Late in the book, Greer finally confronts in frightening detail what is happening in our world now and what will get worse if we humans do not come together to do something. Earlier, Greer had argued that God blessed humans with "free will"—departing from Calvin, who insisted that humans lack free will. The ending of the book is powerful, for Greer gives a call to action, coming back to the idea that we can use our free will to make the right choices and repair the world (or as my Jewish tradition also would say, to perform *Tikkun Olam*).

I am impressed with the seriousness of Alan Greer's endeavor, the immense thought and reading that has gone into writing this book, and the ultimate humanity and generosity of the author. What an accomplishment to work out one's own theology, and to reach a place of comfort that is far from complacent as he urges human action in "partnership" with God.

Achsah Guibbory, Ann Whitney Olin Professor of English,
Barnard College

Preface

God Is seeks to help all of us come to grips with who and what God really is along with what God is not - rather than what we want God to be. As we do so, we cannot ignore reality in favor of our inherited myths; instead, each of us has to take a wide-eyed and open-minded look at all that history, science, and human nature teach us, so that we can achieve a more realistic understanding of the possibilities of God. With such an approach we will begin to see how, if God exists, God works and interacts with us - as well as why God does things the way God seems to do them.

Today there is an ongoing and often heated debate over whether or not God actually exists at all. A significant group of atheists, with eminent scientists such as Richard Dawkins in their forefront, argue that because we and our lives are not perfect, because our physical bodies have "huge design flaws", because bad things happen to us, no few of which are intentionally inflicted on us by other humans in the name of God or religion - and because the human condition generally leaves much to be desired, there can be no God. For them, if God existed, God would be perfect as they want to define perfection and would have made us, as well as all that is around us, perfect. But since the real world portions of this equation are demonstrably not perfect, there can be no God.

In this analysis atheists and non-atheists alike often think of God and argue about God in terms of perfect and perfection based on twenty-first century human definitions of those terms. They then project these human standards onto God and expect God, if God exists, to live up to their current criterions and expectations. This, however, is an evolving phenomenon. For much of human history people viewed

God, in whatever form they conceived of the deity, as being as flawed as they were, just on a much grander scale. It is only in the last two to three thousand years that humanity has begun to try to impose on God the expectation of meeting our definitions of what God's conduct and persona should be.

Being human, people do this because they see and define things in terms and concepts that humans can comprehend. This includes their attempts at an analysis and understanding of God. It has been ever such. Yet today, out of hubris, a number of well-respected individuals have argued that human thinking, knowledge, and understanding have reached such heights that if God exists they should be able to measure God, detect God scientifically and, in fact, judge God.

As part of this, people look back down human history and pity their ancestors for their naive superstitions and approaches to the divine. They even damn them for their benighted ignorance in their pronouncements on these issues.

For example, it's easy enough today for many to shake their heads in consternation at the ignorant thinking of people in the times of Copernicus and Galileo. During the late sixteenth and early seventeenth centuries, these two men disproved the theory that the universe revolved around the Earth. Instead, they showed that our planet revolves around the Sun; nonetheless, for the vast majority of the people living then, it was obvious that the Earth was the center of the universe around which all else revolved - because that's what they thought they saw.

So today it is equally as easy for our contemporaries to think of the then Pope, Clement VIII, as a hidebound example of blind adherence to false and ignorant thinking because he refused to listen to his friend Galileo or look through one of Galileo's telescopes. Instead, that Pope made an Earth-centered universe official church doctrine.

But let's step back for a moment and think about what Clement and the people of his time observed every day of their lives - at what common sense told them had to be the facts. Every morning they saw what appeared to be a rather small but intense ball of light and heat rise in the east and transit the sky above them as they stood firmly fixed on the most immense and stable thing they could imagine, our Earth. They likewise saw the moon and the stars trace their paths across the nighttime skies while the Earth itself sat unmovably in place without their being able to detect its rotation. They thought of distances in terms of how far they could walk, ride, or sail in a day: the equivalent of tens or hundreds of miles. They never conceived of the millions of miles that we now know separates us from our gigantic Sun, which could easily swallow the Earth, but which appears so small in our skies.

Four hundred years ago people's senses serenely confirmed to them that it was the Earth that was the central fixture of the universe with all else moving about it because that's what they saw and felt. They simply could not make the conceptual leap necessary to transition from such thinking to envisioning our planet as a rather small, wobbly, rotating ball of rock circling an immensely larger Sun that, with its planets, was embedded in a nearly infinite universe of other galaxies, suns, and planets.

Without recognizing it, today a host of folks find themselves in a similar position to that of those seventeenth-century people. They can't get their minds and emotions around the concept that they, individually and as a species, aren't uniquely the center of God's universe and attention. Or, for that matter, that God might act or think in ways they don't like or approve of. Our emotional senses and mind set insist that, if God exists, God look at and after each of us in the manner we demand. It is we who understand and feel the pain, joy, and uncertainties of life. Therefore, it is we who should be of paramount importance to God, no matter what we do.

However, a closer examination of reality should teach us that we need to set such thinking and perceptions aside in order to begin to embrace the bigger picture, just as Copernicus and Galileo did. We need to face what are now the proven facts and be willing to recognize their implications. We have to shift our point of view from selfishly looking at the human condition from the inside out to looking at it from the outside in.

People will have to be willing to use all of humanity's knowledge and abilities in order to begin to understand what God is doing, what our relationship with God, and by extension each other, should be, and to achieve some inkling of who and what God is as well as is not. Humanity will have to reach past the easy mindless answers, the answers they have received from the past, and the answers as we want them to be.

Folks will also have to be willing to acknowledge that God is other than perfect. This means that God is especially other than perfect as humans define that term. In saying this we are not implying that God is less than perfect. What we are saying is that God is who and what God is. And that is something that cannot be limited or boxed in by humans in the ways people want to.

Down through the ages some of humanity's very best theologians and thinkers have, one after the other, come to the conclusion that humans cannot truly know or describe God, no matter how hard they try. So far, human language and concepts have just not been up to the task; nonetheless, we can describe attributes and facets of God based on our experiences of life under God and in the universe that surrounds us. We can discern some of the ways God works.

And one of the principal ways God does so is through the physical processes that govern life. These are things people can use their intellects, knowledge, and senses to understand. The history, physics, chemistry, and biology of what God has created are now better known to us. We can see that God works in the real world and universe using its laws to achieve God's will. For example, based on humanity's current knowledge, it becomes obvious that God is in no hurry to complete what began with the creation of the universe billions of years ago, even if humanity is.

With this as a base each of us can try to better understand where God is taking us and why – as well as what God wants us to do with the lessons we are being so patiently taught. It will be that understanding that will allow us to begin to make some sense of our existence and all that makes up the human experience. Then, once we come to something approaching answers to those questions, we will hopefully have a greater understanding and a better concept of that which cannot be captured in either pictures or in words, of God.

Even though as living human beings we cannot truly know who, what, or how God is, we must still try! And in that trying we can begin to dimly perceive the tiniest fraction of at least a part of all that makes up God. And with that merest bit of understanding humanity can grow as God would have us grow. Thus, we cannot be afraid to try because in that trying we also grow. In the words of Immanuel Kant, we "must dare to know."

What we cannot and must not do is to delude ourselves into believing that today we or any one of our faiths, or for that matter all of our faiths together, have the ultimate answers needed to define God or provide the final answers to the eternal questions posed to us. Our search for all that is still ongoing.

God Is is part of this trying, this daring, this searching.

As this book's author that is what I have tried to do in ways that approach these issues from a perspective different from that adopted by many, if not most, of the protagonists in the modern debate as to whether *God Is*. So you, the reader, should understand that though I am a Christian, I am not a trained theologian or a professing member of any denomination. Thus, I am not burdened by an imbibed theology based on any particular group's thinking, dogma, or teachings; therefore, I do not have to implicitly or explicitly defend or adhere to any school of thought. I can, instead, explore the *God Is* issues with much greater freedom, without concern or fear of censure by any peer group that I might offend or have previously committed to.

In my explorations I have drawn on a different experiential background from many currently in the debate I am joining. For starters, I was trained as a marine and electrical engineer and naval officer at the US Naval Academy. Following graduation from that institution I served as an active duty officer for six years. During that time I qualified as a submarine officer and spent my last year of service in Vietnam during that war. After Vietnam I went on to law school at the University of Florida and began what, to date, is an ongoing fifty year career as a trial lawyer.

As such, I have handled literally hundreds and hundreds of complex commercial and political cases. Through each of these I have learned how to take the raw facts which were presented to me, or which I unearthed, and distill them into coherent and logical wholes that could withstand the scrutiny of a trier of fact, be that trier a judge or jury. So this is exactly the approach I have attempted to follow in the ensuing analysis of what I see as the discernable facts that are available to us about God and God's apparent actions.

Additionally, I am a voracious world traveler, reader, and self-taught student of history, politics, theology, and the human condition. A key part of my explorations has been the time I have spent in something approaching fifty different countries and almost all the states and territories of the Union. In the course of all this I've delved into the teachings, practices, and history of most of the world's major religions.

From this base I have tried to look at what the discoverable facts of the human condition and history along with modern science and theology imply about God and our relationship with the divine. You may agree or disagree with this analysis, but at the very least, I hope it will get you thinking about God in new ways, less encumbered by the received ideas that have become barriers for many in their personal quests to reach out to God.

Acknowledgements

I cannot begin to sufficiently express my appreciation to all those who have supported, encouraged and challenged me in my efforts to bring this book to life. But any attempt must start with my literary agent, Lois de la Haba, whose years of diligence to find the right publisher were truly heroic. Right alongside her was her lead editor, Marilyn Meyers. Their editing made this a far better literary work than it would have been without them and I'm so grateful to both.

That publisher turned out to be KCM Publishing and its Publisher, Michael Fabiano along with his team who stepped up to the plate. They have all done such an artfully superb and incredibly thorough job, as well as being joys to work with.

My discussions and arguments over the years with Monsignor John Vaughn and Rabbi Mitch Chefitz have done much to spark and inform the thinking which I later poured into what became this book for your consideration.

And words alone are insufficient to express my appreciation for Achsah Guibbory, Barnard College's esteemed Professor of English who dove in, to craft for your edification as thoughtful a Forward as I have ever read.

It goes without saying that I am deeply indebted to my treasured friends, Dave Barry, David Lawrence, Dr. Joe Greer and Mitch Chefitz for their generous and thoughtful efforts expended in first reading God Is and then writing in support of it.

I also need to gratefully acknowledge the work of Peter Costanzo who has led me by the hand through the world of social media, and my wonderful assistant, Helen Reyes-Vidal who has made my work life far easier than I could hope for.

Finally, and most importantly, I need to thank my law firm Richman Greer, now Day Pitney, LLP, as well as my wife, Patricia Seitz. She not only let me keep my "maiden name," but has both inspired me in this writing and put up with all my idiosyncrasies and maniacal focus on *God Is*.

GOD IS

"I do not feel obligated to believe that the same God who endowed us with sense, reason, and intellect has intended us to forgo their use."

- Galileo Galilei, Letter to the Grand
Duchess Christina (1615)

Chapter One

THE HUMAN CONDITION

To have any hope of understanding even a small part of God, we have to first understand ourselves and the human condition that surrounds us. In contrast to what little we know about God, we know and understand an incredible amount about ourselves. Thus, one of the best ways to better grasp God's relationship with humanity is to look at the human condition itself: the makeup of humankind today and how it functions. We need to do so because, if God exists, God is the God of all of humanity, not just some privileged or religiously correct segments.

As of 2019 there were approximately seven billion human souls crammed onto this lifeboat we call Earth. That, however, is a number hard to fathom in terms of its component parts, so if we proportionally reduce those seven billion people to a hypothetical village of only 100 individuals they would roughly break down as follows[1]:

- 50 would be female, 50 would be male
- 20 would be children
- 80 would be adults, 14 of whom would be 65 years old or older

There would be:

- 60 Asians
- 12 Europeans
- 13 Africans

1

- 8 North Americans
- 6 South Americans
- And 1 person would be from the rest of the world

In terms of nominal religious identification there would be:

- 31 Christians
- 21 Muslims
- 14 Hindus
- 6 Buddhists
- 12 people who practice other religions
- 16 people who would not be aligned with any religion or be atheists

They would be faced with communicating in a myriad of different languages:

- 17 would speak one of the Chinese dialects
- 8 would speak English
- 7 would speak Spanish
- 4 would speak Arabic
- 4 would speak Russian
- 52 would speak other languages

Their levels of affluence would be equally varied:

- 75 would have some regular supply of food and a place to shelter from the weather
- 25 would not have regular access to food
- 1 would be dying of starvation
- 17 would be undernourished
- 15 would be overweight
- 83 would have access to safe drinking water
- 17 of them would have no clean, safe water to drink

In terms of education:

- 82 would be able to read and write
- 18 would be illiterate
- 1 would have a college education
- 1 would own a computer

They'd also have the potential to come from more than 192 different nations. And as of 2010, fifty percent of the world's entire human population lived in urban settings.

These numbers are both revealing and yet, at the same time, deceptive. There is not one Asia, one Africa, one Europe or even one America. In the same way, there is not just one Christianity, one Islam, one Hinduism, or only one Buddhism; instead, the reality is that there is a world of contradiction and complexity lurking under the surface of each.

In Thomas L. Friedman's recent mega-best seller *The World Is Flat,* he paints a picture of a vast globalization in which the world's economy and thus world society has become more and more intertwined, interconnected, and interdependent, yet at the same time hyper-competitive. Isolation for Friedman is a word of the past. But as Ralph Peters points out in his article "The Return of the Tribes"[2]:

"Globalization [only] enthralls and binds together a new aristocracy – the golden crust of the human loaf – [while] the remaining billions who lack the culture and confidence to benefit from 'one world' have begun to erect barriers against the internationalization of their affairs."

In fact much of twenty-first century humanity is retreating into forms of tribalism. In the face of world forces they cannot control or really understand, they have wrapped themselves in their genetic, linguistic, cultural, and religious identities. Zulus and Xhosa begin to clash in South Africa as they jockey for domination, power, and control of a limited economic pie. Catholics and Protestants can't seem to make the leap into a fully governable society in Northern Ireland. Kosovars, Albanians, Macedonians, Bosnians, Serbs, and Croats make a mockery of the concept of European Union. In tiny Belgium those who speak Flemish have trouble finding a way to be part of the same country with the Walloons, or is it the other way round?

In Syria, Yemen, Saudi Arabia, Iraqi, Iran, Afghanistan, and Pakistan there are Sunnis who will commit suicide in order to kill Shias and vice versa, even though both profess to be Muslims. Yet there are those in both groups who will gladly die to wipe out Hindus, many of whom, given the opportunity, will violently attack Muslims of either stripe. Chechens hate Russians who return the favor while at the same time looking down on Ukrainians. Han Chinese clash culturally with Tibetans, Vietnamese, and Uighurs though all of them can be found within China's own borders. Japanese treat Koreans as being fatally inferior while the latter harbor the deepest resentments for the entire nation of Japan. And on and on.

Lest Americans become smug reading this list, in the United States, where everyone, except Native Americans, have ancestors who immigrated to this country within the last four hundred years, and most in far less time than that, chunks of American society loath and fear illegal aliens or "foreigners" who want to follow in the footsteps of most Americans' ancestors.

While humanity lives in one world today, it is far, far from being one world – from being truly globalized. Given this fact, we must ask what are the common denominators that all of these disparate and differing tribes of humanity share? First, of course, they are all human with more or less the same genetic makeups and physiologies. Their physical differences are for the most part on the surface: skin pigmentation, type of hair, shape and color of eyes. The things that make them visually different, one group from another.

Humanity also shares a basic set of wants. All over the world the majority of people want to know that tomorrow, next week, next year

4

they will be able to eat, have a roof over their heads, and be safe from violence. They want order.

In short, people everywhere want the fundamentals: certainty and rules they can understand along with a guarantee that their basic physical needs and safety are going to be met. In opposition to these wants, however, is the stark reality, that at its very core, life and existence are uncertain. None of us know absolutely what tomorrow holds. We can guess, we can predict, but we don't know for sure. And it is surety that people crave.

Thus it is the search for certainty that leads to so much of the negativism and conflict in our world. For example, in her book *A Russian Diary*, Anna Politkovskaya portrays, in depressingly clear detail, the apathy of most of the Russian people towards democracy and personal freedom. Politkovskaya was a highly respected reporter who, in her writings, fearlessly took on public corruption and the destruction of Russia's nascent democracy. She portrayed the horrors of Russia's policies in Chechnya. She railed against the return of her country's police state, which principally benefits Russia's oligarchy and political elites. As a result she was assassinated in an elevator of her Moscow apartment building in the fall of 2006. Hers, however, was a death that produced little in the way of internal public outcry in her own Russia.

What comes across so striking though in her posthumously published *A Russian Diary* is the ordinary Russian's resigned acceptance of all of this. You come to realize that what the majority of Russian society really seems to want is stability and enough to eat. As long as they know the rules, no matter how unfair, and things don't too greatly upset their ordinary lives, they will acquiesce and accept whatever else happens to others both in and out of Russia. They will let a minority of their society manipulate how the rest should live and act as long as their personal world can go on in its ordinary, unchanged, day-to-day way. Even today a major portion of the Russian people continue to want to be told what to do since for them that's the way it has always been.

5

Unfortunately, this same scenario, in multiple variations, tends to be repeated over and over again all over today's world, including in the United States, with only the occasional "Arab Spring" to challenge, for either good or ill, the status quo in spots. All too often the requisite masses of aggrieved people, who must act in concert in order to force the changes required to overcome any given problem, are terribly difficult to muster. More often than not there are just too many forces invested in their accustomed patterns of conduct or current way of doing things no matter how bad the result.

That's because, by and large, no one likes change. There's an apt saying that, "the only person who truly delights in change is a wet baby." Almost everyone else dreads it. Our ingrained human instinct is to resist change. But life is change. Each of us is born, grows up, matures, and then grows old. In our youth we embrace change. We want to get older, get bigger and better. We like the new, the different. But then as people grow older and find it harder to learn new lessons or new ways, they see change as a threat. They resist it out of fear that they can't master it.

Nonetheless, our lives are defined by our need to deal with constant change, both within ourselves and in the world around us. The world's climate changes; our relationships with each other as individuals, social groups, political parties, and nations change. Economies and prevailing technologies change. So one of the most critical things humans do is deal with change.

We don't like it though. It's a threat to most of us. Better the devil we know than the one we don't. We don't want to recognize that it is the very mechanisms of change that have, all over the world, made our lives so much better than they were for our ancestors. We fail to fully acknowledge or understand on the emotional level the day by day, decade by decade, century by century advance of human society through change that has brought us to where humanity is today.

A stark demonstration of this fact can be found in a meandering trip along the waterways of the Netherlands, southern Germany,

Austria, Slovakia, and Hungary. Visiting medieval river towns with their quaint half-timbered houses, magnificent Gothic cathedrals, abbeys, and jaw dropping Baroque palaces, you see how common folks, burgers, prince-bishops, and kings lived in the day they built these edifices hundreds of years ago. But, when you look at the life spans of the notable people who had occupied these structures when they were first built, you realize that the majority of those people were lucky to live into their fifties. And many didn't even achieve that.

Hygiene and medicine were rudimentary. Justice for them was brutal and arbitrary. More often than not it was based on privilege, prejudice, or hate. To be different in any respect was all too frequently a death sentence. Likewise, to be female meant you were little more than a chattel servant and broodmare no matter your social station.

Today, as you see the inhabitants of these same structures, it becomes obvious that, thanks to change, at least those of us in the developed portions of the world live far better and longer lives than the most privileged of people did in bygone ages. Our lives are safer, more rational on the whole, and we have far more opportunities to grow, as opposed to simply surviving long enough to procreate. Thus, over the long haul, change has been our friend, not our dreaded enemy. It is the positive forces of change, whatever their source, that have advanced humanity.

And change will always be with us. It will always assert itself. The status quo is never truly permanent because change is inevitable. So the question becomes, will it be positive change, negative change, or something in between? In the eyes of God and history will it advance the human condition or degrade it? History demonstrates over and over again that when positive change is thwarted, more often than not negative change will take its place.

Along this same geographic tour you can stop in Nuremberg and walk through the history of Nazi Germany from beginning to end: a Germany that felt it had not lost World War I but instead had been cheated, hemmed in, and hamstrung after the 1918 Armistice stopped

7

the fighting and the Allied Powers dictated the terms of the Treaty of Versailles. With a crippled economy and the Great Depression as their spring board, Adolf Hitler and German National Socialism rose to power via "legitimate" elections based on promises of putting down the threat of Communism and a return of Germany to its former grandeur. Once in office Hitler began to systematically obliterate the German people's freedoms, while using the country's identifiable population of Jews and others he considered undesirables as scapegoats for all of Germany's troubles. In short, he was able to achieve massive negative change because the forces of positive change or even the status quo, both within and without Germany, would not stand up to him until the horrors of World War II became inevitable.

On the opposite side of the globe a clique of Japanese militarists and imperialists, whose driving philosophy was the ancient Samurai code of Bushido, hijacked that nation's leap out of feudalism into the industrial age. They turned their sights on the conquest of Asia. To do so they drowned out the voices of those who tried to steer Japan away from military conquest and towards a more peaceful future after their country had essentially been asleep for over two hundred years. During that long night Japan's feudal lords had tried to ensure the stability of their own regimes by freezing Japan in place. In order to achieve this end, they isolated their island nation from what they considered to be all the contaminating change that swirled around it in the rest of the world.

Starting in 1636 a series of feudal dictators known as Shoguns were determined to preserve Japan's culture in what they believed was a pure form. To do so they closed all of Japan's ports to foreigners with the exception of Nagasaki. They next ordered the expulsion or killing of Jesuit missionaries and the wiping out of Christianity as a religion in their islands because it introduced new ideas and beliefs. They also tried to hide from or limit western cultural influences by prohibiting the return of any Japanese who had lived overseas and repressing access to all forms of western thought or literature.

It wasn't until the United States arrogantly forced open Japan's ports through Commodore Mathew C. Perry's 1853 coercive gunboat diplomacy that the Japanese mind was exposed to the changes that were advancing major parts of the world around them. With that event, the gap between Japan's development and that of the rest of civilization became shockingly apparent.

As a result of this long developmental hiatus the Japanese populace was unprepared intellectually or emotionally to resist the negative aspects of change when feudalism was swept away with the opening of their nation to the rest of the world during the 1860s. Popular Japanese sentiment was clearly against this change, this opening of their society to tides that seemed to overwhelm them. In response the general populace rallied around the emblem of Japan's fifteen-year-old Emperor under the slogan "Honor the Emperor and expel the barbarians." This became what is known to history as the Meiji Restoration – a restoration that swept away the repressive control of the Shogun and the feudal lords that had frozen the country in place. This in effect was a revolution that allowed the merchant, industrial and intellectual classes to throw off the Shoguns' stifling torpor that had been holding them back. And it led to Japan's subsequent massive industrialization and its scientific efforts to catch up with the rest of the world – but without its people having had the cultural experiences needed to use these advances peacefully. The end product of all this was Japan's militaristic brand of imperialism that under Emperor Hirohito attempted to conquer the rest of Asia and the Pacific. In the face of this it took a brutal world war and two atomic bombs to shatter Japan's forces of negative change.

Another example of the impact of negative change can be found in the fate of Imperial Russia. At the opening of the twentieth century, Russia sprawled from the Baltic Sea in the west to the Pacific Ocean in the east. Its ruler, the fully autocratic Tsar Nicholas II presided over a hodgepodge of one hundred thirty million people. This population included Russians, Fins, Poles, Slavs, Jews, Germans, Ukrainians,

Armenians, Uzbeks, Tartars, and Mongols among others. Seventy-five percent of these were peasants who'd only been freed from serfdom a generation earlier by Nicholas' grandfather, Tsar Alexander II, and were still principally illiterate.

Nicholas' father, Alexander III, had also been a total autocrat, ruling his country with an iron fist to keep things the way he wanted them to be. He did make advances in the country industrially, building the Trans-Siberian Railroad for example, but despite these limited changes, he kept government and Russian society under the tight control of aristocrats and bureaucrats who were beholding to him alone. Any perceived opposition to his despotic rule was to be crushed out of hand. Censorship was the order of the day.

By contrast, Nicholas was only a pale copy of his father and grandfather. When he became Emperor in 1894 he viewed Russia as being made up of himself as a God-anointed Tsar, the Russian Orthodox Church, and a benighted people. With this mind set he was oblivious to the stirrings of liberal modernism in his land outside of the palaces he moved among. Even Russia's loss of the 1904-05 war with a militarily modernized Japan did not serve as a wake-up call. Things would remain as they always had been with change only occurring when he wanted it to.

Thus, Nicholas and his Russia were woefully ill prepared for World War I, which they stumbled into along with the rest of Europe in 1914. The result was a string of military defeats at the hands of the Germans, horrible numbers of Russian casualties, a crumbling government, and economic breakdown. Despite all this Nicholas' monarchy remained unresponsive to the, by then, churning currents demanding change. The consequences of this benighted response were inevitable. In March 1917 revolution broke out led by the head of Russia's excuse for a legislature, the Duma. This man, Alexander Kerensky, became Prime Minister of a provisional government that tried desperately to solve Russia's problems. At one and the same time it had to battle hunger among the country's masses, prop up Russia' faltering industrial base, remake its society, and still fight the war with Austria and Germany. All the while the government was trying to create a democracy where none had existed before, for a people who lacked both the education and experience to understand or participate in it in any meaningful way.

Russia's German enemy was not going to let any of that happen. In return for a promise to make peace on German terms if he could take power, Germany transported Vladimir Ulyanov, better known today as Lenin, from Switzerland to Russia in a sealed train. And the rest, as they say, is history. Lenin immediately took charge of the Bolsheviks and fomented the overthrow of the provisional government that fell to the November 1917 Bolshevik Revolution. Lenin then established a Soviet state in its place and made a humiliating peace with the Germans in March 1918.

Sudden negative change triumphed. And it did so because the slower forces of positive change, which might have resisted it, had been crushed by Russia's imperial autocracy which hated the very thought of the most important forms of the changes that threatened them.

This phenomenon has not been limited to Russia, Germany, and Japan; instead, it is rampant throughout human history. Just look at South Africa and its white Afrikaner population for example. After their bitter defeat by Great Britain and its colonial empire in the Boer War of 1899 to 1902 in which the Afrikaners, known as Boers, lost control of that area's Cape Colony, the Transvaal, and the Orange Free State, they suddenly found themselves a dispossessed minority in what they viewed as their country. As such they did not have a liking for British imposed governmental forms that would force them to even modestly change their outlooks towards the non-whites they shared that land with; instead, major chunks of this now white minority worked to regain political control and the privileged position they believed they deserved in what had become South Africa. Through dogged resistance they opposed positive change with all their might.

In the face of this, the non-white South African numerical majority struggled to find an effective response. But over the next half-century they were reduced to virtual serfdom even when allied with the liberal leaning portions of the Afrikaner and English-speaking populace who valiantly tried to change the old ways of white thinking. In 1924 a color-bar was introduced in industry. Black African voters were removed from the Cape Province common rolls in 1934 as were the colored voters in 1948.

This was all led by the "Broderbund," a secret Afrikaner organization whose aim was to solidify Afrikaner power and domination.[3] Through the Nationalist Party the Broderbund established pass laws for blacks, limiting their mobility; created tribal homelands within South Africa on the most marginal lands available to foster inter-tribal competition, distrust, and animosity; and generally did all they could to keep the far more vast numbers of South African blacks from ever uniting.

And it worked; for a long time it worked. Finally, however, the black population began to organize and push back. They formed the African National Congress, which was heavily Xhosa, and the Inkatha Freedom Party, principally Zulu. Both organizations resisted through legal means as well as revolutionary activities and "armed struggle." In the end this internal pressure and the weight of world condemnation cracked the Broderbund's white monolithic shell.

Out of all of this emerged a unique leader Nelson Mandela, a Xhosa. Overcoming the bitterness of apartheid and long years of harsh imprisonment, both of which he endured, Mandela saw the possibilities of positive change. He tried to achieve them by uniting South Africa's entire population, no matter their color, into one nation, all moving forward together. But with masses of uneducated or under-educated young people, many of whom are still unemployed; black tribal groups that cling to old tribal thinking; as well as those who seek to consolidate power on tribal bases for just a few, along with endemic corruption the question of whether Mandela's vision will succeed still hangs perilously in the balance.

In the United States there is an interesting parallel to the history of South Africa. Black slavery existed throughout all the American colonies, which were being founded at the same time the Dutch were settling the southern tip of Africa. And slave-based economies became the norm in half the nation. Chaffing at British domination, the American colonies fought a successful revolution. Yet, at the same time, they retained much of the British legal and political systems as well as its thinking, including the enslavement of black Africans. All of this led to a terrible civil war over the issue of slavery, which was fought less than forty years before the South African Boar War.

A defeated South, not unlike the Afrikaners, refused to acknowledge the outcome of that war; instead, institutional structures for segregation of the races and social subjugation of America's blacks were created. Not limited to the southern states, this thinking spread in de facto ways across most of the US, generating barriers for full participation of blacks in American society.

Blacks, in turn, refused to accept their relegation to a life of perpetual poverty and segregation at the bottom of the American social structure, so they formed their own organizations, such as the National Association for the Advancement of Colored People, and began their struggle for full integration, under the law, as American citizens. All of this led to the civil rights movement of the fifties, sixties, and seventies.

Their first significant victories came with the NAACP's 1930 successful effort to block the nomination of Judge John J. Parker to the US Supreme Court based on his openly expressed racial bigotry, the integration of the US military as a result of exemplary black service in America's armed forces in our wars, and in the US Courts in cases such as the landmark US Supreme Court decision of *Brown v. Board of Education*. These successes emboldened larger numbers of blacks. And led by men like Martin Luther King and Thurgood Marshall and women like Rosa Parks, they were ultimately able to force the rest of the nation to begin to confront its own racist mind set. This started the breakdown of all the legal, cultural, and social barriers that had been holding back the advancement of America's blacks and other minorities including Hispanics and Asians. Of course, this is all still a work in progress, but this positive change advanced to the point that the country twice elected Barack Obama as its first African American president.

Out of all this, it becomes obvious that in human affairs the one sure thing is change, either predominantly positive or negative in nature. The corollary to this rule is that when positive change is thwarted negative change will more often than not take its place. No matter how hard we humans try, we cannot enforce the status quo as a permanent condition of our existence. Other forces will always push us towards some form of change. Be those forces population explosions, changing economic conditions, new technology, climate change, ideologies, attitudes, the search for resources, too much or too little water, or other disasters: pick your poison. Something will always crop up to force human change for either good or ill.

Nonetheless, for most of us, so long as our personal condition is more or less OK or even just tolerable, we tend to want to be left alone to either live with things as they are or to try to make them better for ourselves individually on our own terms. Most people don't want to be asked to pay more taxes to help people we never met, or to have to deal with problems that we can't touch, feel, or see up close and personal. That is until uncontrollable change overruns us, and we scream for help, human or divine.

In fact we have recently lived through a prime example of just such an event. As part of humanity's evermore challenging search for fossil fuel resources, we are drilling for crude oil and natural gas reserves buried under tens of thousands of feet of the earth's crust that in turn lie deep beneath thousands of feet, if not miles of ocean. At one of these drill sites in the Gulf of Mexico known as the Deep Water Horizon Platform, the British Petroleum Corporation, or simply BP to most of us, was drilling a well on the seabed more than five thousand feet below the surface. And on April 20, 2010, disaster struck. There was a blowout at the submerged wellhead and a resultant fire on the drilling platform that left eleven platform workers dead. The well-head's mechanical shutdown system, or blowout preventer, failed to live up to its name and for nearly three months a total of more than two hundred million barrels of crude oil gushed into the Gulf's waters in what was euphemistically referred to as a "spill."

But this wasn't a glass of liquid that had been knocked off your table and spilled onto a clean floor. It was and is a monumental disaster: one that shut down vast areas of the Gulf's marine fisheries as well as contaminated thousands of square miles of beaches, estuaries, and marshlands essential to the environmental chain of life in the Gulf. Recreational tourism and the economies on the coasts of four states went from vibrant industries to the equivalent of ghost towns in days.

The people who live and work on those beaches, coasts, and estuary systems are, by and large, hardworking, religious, and often deeply conservative. They are good people who like to think they can take care of themselves. In parallel with all this, they are also innately distrustful of national governmental policies that they see as imposing wastefully unnecessary taxes and regulations on their ways of life, when they just want to be left alone.

Nonetheless, in the face of such a disaster those same people suddenly needed and wanted help from anywhere and everywhere. Send in the EPA and the Coast Guard. Make BP pay. Washington, God, somebody do something to change what was happening to them. Make everything go back to the way it was. They desperately wanted the government they used to denigrate to suddenly show up with a magic wand and return their lives to the way they were. And if government couldn't do that, then people really thought maybe God should.

They were faced with change, in this case very sudden, unwanted, and cataclysmically negative change. In the face of that change, national, state, and local governments along with BP, the corporate culprit, all swung into action with a massive containment, shutdown, and cleanup effort. One of the key results was that after agonizing months

15

of gushing incredible amounts of crude into the Gulf's waters, the Deep Water Horizon well was finally plugged.

The incredibly complex efforts necessary to do all this could only be accomplished because decades of positive change had created the scientific and engineering knowledge, industrial strength, and governmental organizations necessary to mounting such an effort. But the change the peoples of the Gulf coast had never wanted had still been visited upon them. And they will be both paying for and dealing with it for decades to come.

During all this, the rest of America and the world watched in a detached sort of horror. Wasn't it terrible people asked as they drove up to gas pumps and looked to see if the prices they'd have to pay had gone up. Outside of the green parties there was no mass movement for safer drilling practices or a cry that the country should be weaning itself off such dearly gained oil. Like the people of the Gulf Coast, the rest of the nation didn't want their accustomed ways of life interrupted by such unpleasantness. For a while most stopped saying, "Drill, baby, drill," but only for a while. America is still hooked on that oil. It is still adverse to that much change.

True, people want their problems as well as the world's problems solved; however, by and large folks want someone else, some other agent or force, to do it. They want instant government action by faceless others to pass a law, issue an order, give America an easy solution. Absent that, people want magic or miracles but not ones that impose change on them individually.

As part of all this, many invoke religious or supernatural beliefs in the hope that they will provide the answers everyone wanted. And they do it in the same way people have been doing since the earliest days of humanity. They and we have felt that there were powers surrounding us which controlled life's events that went beyond ordinary natural forces. We humans have been calling on those powers for tens of thousands of years to provide the solutions or answers our own minds and muscle couldn't supply.

Yet all too often when humans invoke supernatural forces they have wanted to not only benefit from them, but to, in fact, dominate them as well. Just think of the classic fairy tale of Aladdin and his magic lamp. Humans secretly want to control the lamp's genie – to be able to force him to grant their wishes: in short, to be able to command the supernatural.

In fact, domination and control are two of the forces paired with change. They are intertwined with our search for ways to deal with change and our need for fixes to humanity's perceived problems. This drive for domination is one of our fundamental instincts. We seek to dominate our environment as well as anyone or anything else that would impose change on us.

Whether it is out of a thirst for power, ideology, religion, fear, survival, or just the basic animal impulse to be the alpha, either individually or collectively, our fundamental human instinct is to reach for, or avoid, domination. Thus there is a constant tension in all our relations with each other; or our culture, government, and religions as to who or what is dominant. The sagas of Nazi Germany, Imperial Japan, revolutionary Russia, apartheid South Africa, American segregation, and the Gulf oil spill are all prime examples of our human drive to dominate or avoid domination.

In doing this, people even seek to dominate God, the divine, the supernatural. They want those forces to be who and what they think and believe they should be. They try to make them do what they want done: cure their ills, crush their enemies, damn those they despise or look down on, advance their cause. How long do you want the list to be?

Currently the Taliban and the Islamic State in the Levant or, alternatively, in Syria (ISIL, or ISIS if you prefer) are probably two of the most visible examples of the drive to dominate in order to stifle positive advancement and impose retrograde negative change in its place. Invoking Allah and the Koran, Afghanistan's Taliban and Syria's ISIS seek to impose a rigid interpretation of Islamic life and law as it was practiced in, say, the eleventh century on modern day Afghanistan and

17

the Middle East. They seek to force females, and by extension society itself, back into burkas and out of schools because they are unable to cope with or compete in a twenty-first century world where women are educated, openly work as men's equals, and have individual and political rights, and where the Koran is often interpreted in ways that accommodate the complexities of post-industrial societies. They do so in order to keep women and others they despise subservient and non-competitively passive. They try to ban all forms of what they perceive as ungodly western culture. And they abhor anything that smacks of beliefs, even Islamic beliefs, that differ from their own.

Both ISIS and the Taliban are more than willing to kill and maim anyone who gets in their way as part of their tactics of intimidation. Theirs is a "holy war," a jihad against the rest of the world if that world comes to Afghanistan or ISIS's "Caliphate." They cannot conceive that Allah, God, could ever accept change in the human condition subsequent to what it was at the time of God's dictation of the Koran to Mohammed, his prophet in the early 600s C.E.

But what they are really trying to do is to impose their will on both Allah and the people surrounding them, to dominate both. By their actions, what they are saying to God is let us tell you how things should be and how we will make them be. They want to ignore everything that has transpired outside of the Islamic world since the seventh-century life of Mohammed as something God must or should not have wanted. As such they are also saying that God could not control or direct those non-Islamic events, so they, the Taliban and ISIS, will show Allah how to put things right. In essence they want to dictate to God, to dominate the deity.

Thus ISIS and the Taliban are the perfect examples of a constantly repeated trait in the human condition. The one in which people all too often believe that what is important to them must also be important to God in the same way and to the same degree as it is to the believer. So if goats, sheep, and cattle are important to humans, then God must equally value those same animals and be pleased to receive them as sacrifices as well. If they are afraid of women or blacks who can compete with them intellectually and in the job market, then they say God must want those same blacks and women consigned to a place and condition where they can't so compete. If they don't like change, then God must not like it either. In this way they constantly project their thinking, values, and prejudices onto the divine. They want God to be perfect – but perfect as they define perfection.

In doing this many such people ignore the grim challenges facing the rest of the human family because, in their heart of hearts, they want to believe God must want those other people to live in such conditions. Or, alternatively, they say let God fix those other problems if God wants them fixed. Just don't suggest God is asking them to personally change their own ways of thinking or beliefs. It is this attitude that, if not checked or resisted, creates negative change for humanity.

In this light the question becomes, if God exists, does God intervene and influence the processes of positive change leading us past negativism. Or does this all just happen out of chance and human nature?

Chapter Two

WHAT CAN WE DISCERN ABOUT GOD

What does this picture of the human condition suggest to us about the God so many people want to dictate to, as well as the question of God's existence? Through the ages priests, theologians, and everyday people have struggled to understand God and our relationship with the deity. They have dreamed of knowing God, but time and again they have run into mental or emotional brick walls in their efforts and thrown up their hands in frustration. The consensus has been that, if God does exist, God is a mystery and unknowable.

In the thirteenth century Thomas Aquinas wrote, "This is what is ultimate in the human knowledge of God: to know that we do not know." He elsewhere said, "…the human mind…does not reach a knowledge of what God is, but only that he is."[4] Some seven hundred years later the twentieth-century Jesuit philosopher and theologian Karl Rahner opined in the same vein, "I must confess to you in all honesty that for me God is and has always been absolute mystery. I do not understand what God is; no one can. We have intimations, inklings; we make faltering, inadequate attempts to put mystery into words. But there is no word for it, no sentence for it." He goes on in another passage to say, "The task of the theologian is to explain everything through God and to explain God as unexplainable."[5]

The vast majority of us would agree that God is ultimately unknowable. But that is not the same as saying humans can't know or discern some things about God and how God works. That we can do. We can come to understand at least something about God by looking at God's choices and the methods God has elected to use. We can then try

20

to analyze why God did and does what God does as well as the implications that arise from all this in terms of its impact on humanity.

But to begin any attempt at understanding what God is doing and God's relationship to humanity, we need to try to simultaneously view God's conduct from multiple vantage points. This requires looking at what is known about God's orchestration of the universe, our planet, life, and humanity; as well as our own individual lives from their inception to today.

One way of doing this is to think of all these interrelated pieces as a great and endless ballet. One in which the audience is first drawn to the individual parts, viewing them up close and personal as you would the individual dancers - with all their passion, virtuosity, and flaws. They need to be aware of the new performers flowing out onto the stage even as other dancers leave it. Then, simultaneously, the audience must look at the overall dance observed from a distance and from above in order to be able to see the intricacies of its patterns in motion as they unfold across time.

While doing this the viewers must try to decipher how, under God's direction and teaching, those interlocking patterns and performances have evolved over the eons as well as where they seem to be going – what are they pointed towards. What will they morph into in the future? At the same time viewers need to remind themselves that God is both the choreographer of this dance as well as the ballet master who taught the dancers the steps God expects them to execute.

At the same time the audience also needs to keep in mind that it is up to each dancer, individually and collectively, to perform as they will to the best of their abilities. Finally, each member of the audience must attempt to unravel the meaning of the dance itself as well as the message that meaning conveys. Then, ultimately, each must ask what all this portends for our future?

So assuming that God exists, what can we humans learn or understand about God from the current human condition and the facts, scientific and otherwise, that can be observed or are generally accepted as proven? First and foremost, God is not the God of only we humans or even just our Earth.

God created the entire universe. This includes both the observable universe as well as that beyond our powers of observation. Turning to just that portion of it that we can detect from here on Earth we know that at the very least it stretches away more than thirteen billion light years in all directions.

Within that sphere of observable space there are well over one hundred billion galaxies not unlike our own Milky Way galaxy. Our galaxy just by itself contains over two hundred billion individual stars, each a unique sun. And if we could travel across thirteen billion light years of galactic space in any direction to the edge of what can currently be observed of that space, there would probably be another thirteen billion years of visible universe spread out in front of us. That new horizon of space, which we cannot presently detect from here on Earth, would probably be equally filled with billions of more galaxies. In fact it could stretch on to infinity.

We now know that this entire mind-numbing vastness burst into existence 13.8 billion years ago in what is currently referred to as the Big Bang. Again, assuming that God was the author of this event and everything that has proceeded from it, including the creation and recreation of endless numbers of stars, planets, and galaxies as well as all that lies between these bodies (more about which we will discuss in later chapters), we come to the birth of our Sun and its solar system including Earth some 4.5 billion years ago. As part of that creation our Earth's orbit came to rest at the exactly right distance from the Sun, sometimes referred to as being in the "Goldilocks zone," in which it was physically possible for our planet to serve as a cradle for life.

From all this, a number of primary facts can be deduced about God. Chief among these is that God is patient, incredibly patient, and perhaps even infinitely patient. As a corollary to that patience it would appear that, given the 13.8 billion years God has spent shepherding forward our universe and all the life it must contain, God is obviously in no hurry. But then God would seem to have no need to be. This timeline suggests that God has eternity to work with.

It would thus also appear that God is an eternal being. God must likewise be an infinite being[6] stretching across, encompassing, and permeating not only the incredibly vast universe just described, but all the rest of the other parallel universes, if they exist, that have been postulated by a number of cosmologists; likewise, God would be the God of all the extra dimensions mathematicians and physicists are currently playing with. By the same token, if an extra-dimensional heaven exists, God must cover all that as well. So God is both infinite and eternal. In short, God is the God of all creation forever.

In these terms, one simplistic way of conceptualizing God is as cohesive sentient energy capable of not only thought, caring, love, and action but also all the rest of the universe of emotions, actions, and intellectual activities that go beyond mere love, thought, and caring. In saying this, in thinking of God as a being of energy, it is helpful to consider what the cosmologist, Dr. Paul Davis wrote in his book, *The Goldilocks Enigma*[7] in reference to Einstein's famous equation $E=mc^2$. That equation tells us that "mass is energy and energy has mass." Thus, God as energy would have substance and not be ephemeral or a figment of the human imagination. In fact, in the Bible Jesus teaches that God is spirit,[8] which is about as close as the age in which he spoke could come to the concept of pure sentient energy.

This concept of God is far different from the way most modern-day believers mentally conceive of God. We are taught that humans are created in God's image, so many people think God must therefore look like us. Our concepts of the deity are influenced by manmade images of God such as the larger than life sort of gray-bearded old prophet so beautifully depicted by Michelangelo on the vaulted ceiling of the Vatican's Sistine Chapel.

Our thinking is also influenced by Biblical descriptions of a divine being sitting on God's heavenly thrown. This suggests a human-like form and actions we can understand. In this same way Christians proclaim that Jesus is the "son" of God. So it's easy to translate this into human terms and surmise that a biological son must have a physical resemblance to his father. Thus, God the Father must have a physical shape not unlike Jesus the son.

In this way, as part of the human condition people project their human concepts and thinking onto God. But the capabilities and reach needed to instantly encompass and permeate the endless vastness of the entire universe, while simultaneously being aware of all that is part of and occurring in that universe, cannot be crammed into one oversized prophet. No possible version of the human form, no matter how gigantic, could ever reach the levels of an all-encompassing, sentient, living presence spanning the entire cosmos that God is and has to be; instead, God is something far different from the mental picture of God so many of us have built up in our minds.

That said, however, simply throwing up one's hands and declaring God to be unknowable threatens to turn God for many into a distant, unfathomable, and capricious deity who is not to be trusted. God is certainly not that.

Therefore a key component of any effort to achieve the beginnings of a basic understanding of God requires a more measured approach. We need to be brutally wide-eyed and realistic in our analysis of how God works in our world. The good news is that, as a species, humans are now finally mature enough to do so; however, as part of doing this each of us must stop trying, as most of us do, to force God into the mold and form of what we want. It does no good to say God cannot be God unless God does or is X or Y. God is God. As the one and only deity, God is what God is and does what God has determined serves God's purposes, not ours.

One of the ways people can approach such an honest analysis is by looking at how God apparently operates. Another is by trying to discern God's characteristics – what God wants and is trying to accomplish.

All that is not necessarily as difficult as you might at first be tempted to think. So let's begin this effort, this analysis.

Chapter Three

WHAT SUGGESTS THERE IS A CREATOR

As part of our attempt to begin to understand something about God, we first need to look at what God has done up to this point with and in our universe, as well as with life here on Earth. Included in this analysis is the question of how God has accomplished all this incredible work. These will be our initial signposts, so we need to begin at the beginning. In doing this we must strive to keep an open mind, while we seek a better understanding of the facts that can provide us with some of the answers we are looking for.

Our starting point in this endeavor would be the moment before the Big Bang, the event that set into motion everything that has led to us. That's where existence itself changed from there being essentially nothing to something. One of the two key words for all of our analysis in this endeavor is change. The other is consistency. These two words should be our basic guiding stars in discerning what God has done in the real world. As we will see, God works through large and small changes over vast periods of time, all the while keeping certain primary factors immutable. It is through incremental change coupled with a set of permanently fixed values in the realms of physics which values are ordained by God that God has produced our physical universe.

This universe we are focused on is, however, an utterly amazing place, and the same can be said for our planet, Earth. The universe's governing laws – those of quantum mechanics, or as some physicists refer to them, quantum physics, Newtonian physics, chemistry, biology, and mathematics – are so consistent and precisely formulated that they produced our universe in all its vastness, our planet, and life as we know it. Even small variations in any one of an incredible number of the component parts or sequences of these laws would render our universe and us impossibilities. If just one thing had gone wrong or was different in the process and its chain of positive events, no universe friendly to biological life, no us.

The issue then becomes what or who ensured that all the necessary factors and variables in the chain of events that constitute this progression would turn out to be just right for us? That chain, by the way, starts with the first nanosecond of time after the Big Bang at the most basic subatomic molecular levels. It then proceeded forward in a precise sequential order all the way up to the universe itself and us. If anything had gotten out of place in that sequence or broke this chain, poof, we and the universe we depend on wouldn't be here.

So was it God or chance? To make up our minds as to how to choose between the two possible answers to that question, chance or God, let's look at a summary of just a portion of the more understandable of the interlocking cosmic lottery wins that had to come into existence in just the right sequences for us to be here reading this. Obviously we have to start with the Big Bang itself. That event occurred approximately 13.8 billion years ago. The instant before nothing existed. Not matter, energy, or forces. There was no universe. Time itself had not begun.

But with that unbelievably gigantic explosion which came out of nothing, all the makings of our universe burst into existence. When we say the makings, we are talking about all the particles of matter and forces needed to create, over billions of years of time, the immense expanse of galaxies, stars, and planets that now populate the universe.

The driving forces that both created and ejected all those particles outward from nothing – as part of that initial explosion – had to be exactly balanced against the force of gravity, which winked into existence at the same time. If the outward explosive push had been too strong, the universe-wide plasma field of matter particles that burst into existence would have dispersed like gas in a void. If, on the other hand, the countervailing gravitational forces that likewise suddenly appeared and began pulling against that expansion had been greater than they were, that plasma would have collapsed back in on itself under gravity's pull. Either way, there would have been no material universe as we now observe it in the heavens above us.

And keep in mind that in almost 14 billion years that bang has been a onetime only event. That is to say it has happened just once and has not been repeated since then. If it had, everything in existence would have been annihilated in the searing heat and radiation of that next explosion and the whole process would have had to start all over again. But that hasn't happened.

As it was, however, the resultant particle plasma, which was so suddenly created out of nothing, was made up of vast amounts of both matter and antimatter. We know from our science classes or sci-fi movies that when matter and antimatter collide, they annihilate each other. If that primordial plasma of subatomic particles had been equally balanced between antimatter and matter, it would have destroyed itself leaving nothing but a massive field of radiation. But that wasn't the case; instead, there was an infinitesimally small imbalance in favor of matter over antimatter. This, of course, meant that after all the antimatter had finished off itself and a corresponding amount of matter, what was left was enough subatomic matter particles to create our universe of nearly endless numbers of galaxies, suns, and planets. However, it didn't have to be that way. That original nearly endless cloud of particles could just as easily have been equally balanced between matter and antimatter, or even have had an imbalance that favored antimatter. In either of those cases no universe, no Earth, and no us.

The remaining subatomic particles of matter that survived had to also be just right in terms of the infinitesimally small, but nonetheless real, amounts of mass and electronic charges each particle carried in varying degrees. Without the correct combinations of mass and electrical charges the remaining subatomic particles could never have combined into the various different atoms that now make up the stuff of the visible universe we see around us. And once again those electron and mass-per-particle properties could well have been different resulting in no hydrogen, oxygen, carbon, iron, etc., that we are dependent on for our existence.

But in order for these vast amounts of matter particles to, in fact, combine into the critical atoms we depend on, two other things, two forces, had to come into existence calibrated with just the correct strengths and ability to interact. Those two forces are known today as the strong nuclear force and the weak nuclear force.

The strong nuclear force is what binds particles together to form the atoms we and everything around us are made of. The weak nuclear force in turn allows some of those atoms to undergo a species of synthesis as they morph into the other atomic forms of the rest of the elements that constitute visible matter. If either of these two forces had been slightly stronger or weaker, our atoms wouldn't have been able to form or hold together as the solid world we experience today.

Initially, those two forces combined to create just a few different types of atoms: hydrogen and helium making up the majority of them. Today those two elements constitute ninety-nine percent of all the matter we can directly detect in our universe. It is the synthesis of the other one percent of those atoms under the influence of the weak nuclear force that has produced all the rest of the elements such as carbon, oxygen, and iron that are so vital to our existence. And once again, that synthesis shouldn't have happened absent a set of extraordinary circumstances. Something had to jimmy the system in order for these elements to appear in the amounts needed to make our universe biologically friendly. And it did.

Under the normal laws of nuclear physics we should hit a bottleneck at carbon, the fifth element to form following hydrogen and the sixth in the periodic table of elements. Given carbon's composition of six protons and six neutrons it should have been extraordinarily difficult for the strong and weak nuclear forces to combine the necessary sub-atomic particles in such a way as to produce the vast amounts of carbon we find in the universe.

Incredibly, at just this point an unexpected resonance appears. As a result of that resonance atoms of helium, the second element, could interact with the forth element to form, beryllium producing carbon, the sixth element. This resonance only appears at just the right energy level setting of the strong nuclear force. If that force setting had varied by less than one percent the binding energies of atom nuclei would change to such an extent that the formation of carbon in any quantity would have been impossible, and without carbon, no carbon-based life forms, us. But, with this bottleneck removed the synthesizing process could move on to the creations of the rest of the elements that make up our visible universe.

Taking a step back, however, something else happened at the formation of those vast clouds of hydrogen and helium. With their appearance the forces of gravity continued their work in a different way. The gravitational pulls on all those atoms interacting together and in conjunction with the gravitational forces of dark matter (which we will discuss more fully a little later on) caused them to begin aggregating into larger and larger galactic clusters, which over millennia eventually coalesced into numberless individual star systems grouped together in protogalaxies. When each of these protostar balls of hydrogen achieved a sufficient density, its resultant critical mass generated gigantic thermonuclear explosions, which in turn burst into star after star in endless numbers.

And once again the setting of the force of gravity played another critical role in this process. Its consistent strength across the entire universe controlled and dictated the rate at which each newly ignited star burned up its accumulated fuel source, the amount of hydrogen it was made of. This in turn determined how long each star would last. Using our own star, the Sun, as an example, if gravity's strength were, say, doubled, our Sun's lifetime would have shrunk from about ten billion years to a mere 100 million years. Given the fact that it has taken 4.5 billion years for life to evolve to its current state here on Earth, a sun that lasted less than that long before it burned out wouldn't have cut it, so something had to, once again, cause gravity's strength setting to be just right and it did.

Chapter Four

EARTH

Since the Big Bang, this process of galaxy and star formation has rolled forward and outward, creating more and more stars even as many of the earlier stars either burned out or, if they were massive enough, exploded as supernovas. It is those supernovas that from time to time illuminate our night skies with a far brighter pinpoint of light than that of the other stars. And each such exploding star acted as a sort of gigantic blast furnace to create the heavier elements needed to form our Earth and the other planetary systems spread across the universe.

Thus, about 4.5 billion years ago, enough matter, including huge amounts of hydrogen, were drawn by their combined gravitation attractions into a swirling mass of rotation that coalesced into our Sun and its solar system of eight planets. This set the stage for humanity's later appearance in the universe.

The third of these planets outward from the Sun, our Earth, turned out to be in exactly the right circular orbit around the Sun and at the correct distance from it to allow liquid water to exist on a year-round basis, an orbit in what is referred to as the Goldilocks zone for supporting life. It also had sufficient mass to create a gravitational field strong enough to hold in place a breathable atmosphere once that developed. Its mass was likewise composed of all the elements needed to support biological life and human civilization, including carbon, oxygen, water, and the other heavier elements. Within this Goldilocks zone the Sun's radiation produced surface temperatures on Earth that were neither too hot nor too cold, but instead they were just right for life to evolve. And Earth's twenty-four hour period of daily rotation also allowed this radiated heat to spread more or less evenly over our planet's entire surface on a year-round basis.

One of the critical components of our planet's mass is its enormous molten iron core. As part of Earth's axial rotation, that core spins at such a high rate of speed (about a thousand miles an hour at the equator) that it generates an Earth-enveloping magnetic field known as our magnetosphere. This field shrouds our planet and us with an invisible electronic shield that forces a majority of the solar winds blasting outward from the Sun's surface to bend around Earth as they blow by us. Without this shield to deflect those winds of solar radiation, they would strike our planet full force and fry all life on Earth to a crisp.

None of these factors were preordained by nature to be as critically balanced as they are. Several of our companion planets have more elliptical orbits around the Sun. If Earth's orbit were of the same elliptical shape, our planet would have wandered in and out of the Goldilocks zone preventing biological life from evolving here for, among other reasons, lack of year-round liquid water.

Likewise, other of our solar system's planets have axial rotation periods of a hundred plus Earth days or more for a single "daily" planetary revolution. Just think what that sort of slow rotation would have done to our evolution if Earth turned on its axis at a similarly languid pace – months of endless night and freezing temperatures on the side of Earth facing away from the Sun followed by months of blistering heat when that side finally rotated back toward the Sun. The result would have been far, far less hospitable conditions for biological life. Under those circumstances, life as we know it would probably never have come into exist.

Another critical factor is the composition of the overall mass of our Earth, which in the Goldilocks zone sense again turned out to be just right. If Earth had the mass and elemental composition of say, Mars, our next-door neighbor, our planet's gravitational field would have been too small to retain the breathable atmosphere needed to sustain complex life forms just as Mars does not. If our planet didn't have an iron core, as many planets don't, there would have been no life-protecting magnetosphere to shield us from the effects of the full blast of the Sun's solar winds.

Thus, everything here on Earth had to have just the right combination of factors, placement, and properties to permit biological life to exist and evolve. And it does, when it could easily have been far more hostile and different; likewise, all this just rightness is part of the far longer chain of factors we've previously outlined, all of which had to be exquisitely calibrated in order to produce us and all the rest of the life here on Earth.

Chapter Five

THAT AMAZING CELL

But what about life? Where did it come from? How did it spring into existence? All life flows from one common ancestor cell; a cell that was the very first living thing on Earth, which we will call the primary parent cell.

Some 3.5 billion years ago when our planet was finally ready for it, that first amazing cell appeared as if out of nowhere. And to say amazing is a vast understatement because where before there had only been inanimate matter there was suddenly life, incredibly complex life wrapped up in one single living cell. The divide, the vast gulf that separated inanimate matter in its most complex forms and the living complexity of that first minute cell is hard to describe. It was as if you had to leap across an entire ocean in one jump.

Obviously we have no fossil evidence to tell us what that very first parent cell looked like, but based on the enormous advances of the last fifty years in molecular and cell biology, we can make some very good educated guesses about its makeup.

In the first place we are talking about something so small that hundreds of them would take up less space than the period at the end of this sentence. In fact, to look at cells today we need very good optical microscopes just to see them, and to peer into their internal structures requires the very best electron microscopes available to science.

Yet that first minute cell contained a surprising array of components. It was also a veritable micro-miniature computer that could and did transmit abstract intelligible instructions to its internal operating parts and offspring cells.

Based on what we know from the myriads of studies that have been done in cell biology, "we must suppose that it [the parent cell] contained a prototype of the universal machinery of all life on earth today."[9] This includes every blade of grass, leaf, flower, tree, gill, fin, nervous system, blood cell, internal organ, and brain found in any form of earthly life. Its cellular structure thus provided a template for all the living cells that have ever existed since then, including ours. It would seem that, while wonderfully complex, the basic internal structures and chemistry of all cells across all forms of life are essentially the same and have been so as far back in time as we have any evidence or suspicion, starting with that first parent cell.

The primary parent cell would have been encased in a membrane or cell wall that was at the same time tough yet porous. This membrane would have in turn enclosed a plasma membrane containing its cytoplasm and individual genome or DNA. Like its entire offspring line this primary parent cell would have absorbed nutrients through those walls to generate food energy and then expelled its waste products back out the same way.

The cytoplasm in turn can be visualized as a liquid gel whose molecules produced many of the chemical activities that are fundamental to its living existence, including the breakdown of the nutrients it had absorbed to create its energy. Those gel molecules also dictated the movements within and without its cellular structure, and its production of proteins that were vital to its growth.

And that protein growth would lead to the parent cell's division into two cells, the parent and the child cells, followed by their further division into more and more cells all reproducing through this process of cellular division. As part of that division each of those cells had to exactly duplicate the prior cell's entire structure, contents, and capabilities. Once it did so the new cell would break off from its parent cell to begin its independent life repeating this entire process.[10] In essence each cell gave live birth to a new daughter cell.

35

It is reasonable to assume that the primary parent cell was some form of microbe or bacteria, the simplest forms of life known to science. They can exist in the most hostile of environments including concentrated brine, the acid of volcanic springs and beneath frozen Antarctic ice, and they can repeat their processes of cell division at prodigious rates – once every twenty minutes or so. Bacteria, however, lack the cell nucleus and many of the subparts of the later vastly more complex cells that evolved out of the parent cell.

To do all this, the primary parent cell needed one more thing to make it all pop into existence, something that had never before existed in the previously inanimate world. And that was the equivalent of a biological computer to organize and direct all this living activity. This was a genome made up of DNA that acted as a blueprint for the life that was to come. So where did that genome come from? How did it suddenly come into existence?

To put this question into prospective we can look at one of the most studied of bacteria *Escherichia cola* better known as *E.coli* for short. Its genetic code or genome, the chemical code that defines its shape, structure, and what it can do is about four million characters long. If printed out it would fill at least one thousand pages of a book. The human genome by contrast is over 3.5 billion characters long and would fill a library.[11]

A genome consists of the chemical equivalent of computer algorithms that tell the cell what to become, what to do, and how to do it in the same way your personal computer operates and preforms the various functions you use it for such as word processing and reading the information you store on it. Those chemical algorithms in turn are made up of three letter "words" with each of those three letters taken from a chemical genetic alphabet. The chemical algorithms constructed by these genetic words have to be precisely "programed" with each of its "words" in the exactly correct sequence in order for them to have meaning and convey the abstract information and commands needed to create and direct the cell's activities and life.

So how did this genetic alphabet come into existence? And what or who wrote the words and then the algorithms made up from them? What caused all these millions of words to fall into the precisely correct order needed to create the first living parent cell's genome?

Those who wish to deny the possibility that there was any sort of intent or intelligence behind this design of what became the primary parent cell's four-million-character genome argue that it all happened through some long sequences of chance occurrences. Just another in the already incredibly perfect sequences of chances that we have been looking at in earlier chapters.

But when you think about it, that's like saying that you could take every single word and punctuation mark in this book apart, letter by letter, mark by mark, and then dump them all into a sack, stir, and thoroughly shake all those characters up. Once they are completely mixed, upend the sack and pour them out bit by bit onto page after page of paper and expect them to arrange themselves by sheer chance back into this same book with all its words, thoughts, and arguments in exactly the sequence you are now reading. It defies credulity to even suggest that could happen, but in essence that is what those who argue for the chance creation of the parent cell's genome are proposing.

And this sets aside the question of where chemical DNA and DNA's alphabet came from in the first place. How did it come into existence if it was not designed by a creator? How could chance have produced the chemical alphabet of life that would be capable of being used by a genome through accidental combinations to convey the intangible information and instructions needed to turn inanimate materials into living cells all at once and at the microscopic levels of cells?

Chapter Six

—————◆—————

EVOLUTION

Now let's look at that biological life itself. We begin by once again going back in time to the formation of Earth 4.5 billion years ago, or just over 9 billion years after the Big Bang. Over the next billion-year period Earth, riding in its Goldilocks orbit, began to solidify into a grouping of land formations surrounded by primordial oceans of liquefied water. Within those seas a complex set of chemical reactions and changes were initiated that led to the creation of proteins that could chemically reproduce themselves. This went on until finally about 3.5 billion years ago, in another one-time-only event, the first minute living cell, probably some form of microbe, emerged from that soup of reproducing protein water. This one single cell then proceeded to successfully replicate itself over and over again.

What makes us think this only happened once? The answer is the common DNA genetic code found in all living things today. Be it plant or animal, all across Earth's land masses and in all its oceans, all its life is based on the same immutable DNA code. That genetic code has remained fixed and unchanged over the last 3.5 billion years. And it all comes from that very first primitive living cell that managed to survive and reproduce itself.

But why just one DNA code? Why not multiple differing genetic codes? Assuming that the bands of ocean waters circling our globe all those billions of years ago had relatively similar conditions of water temperature and chemical makeup over vast stretches of those primordial seas, shouldn't they have produced parallel events of cell evolution at multiple geographic locations? If they had, at least some of

38

those primitive cells should have had dissimilar versions of a DNA code. Such a multiplicity of events would have led to differing lines of DNA based evolution. And those differing lines should have produced dominant creatures with substantially different characteristics – such as, say, six limbs and eight eyes or other significant differences from the four-limbed, two-eyed variants we find so prevalent today.

That didn't happen, however. Instead there was and is just our one common ancestral DNA gift that has come down all the living lines of biological life to us today ranging from the minutest microbes to the largest living whales and all the universe of plant and animal life in between.

As the biologist, Richard Dawkins (no friend of the hypostasis of God's existence) puts it in his book *The Greatest Show on Earth*: The Evidence for Evolution:

"...the genetic code is universal, all but identical across animals, plants, fungi, bacteria, archaea and viruses. The 64-word dictionary by which three-letter DNA words are translated into twenty amino acids and one punctuation mark which means 'start reading here' and 'stop reading here' is the same 64-word dictionary wherever you look in the living kingdom....[This] genetic code is a 'frozen accident' which once in place was difficult or impossible to change....Any mutation in the genetic code itself (as opposed to mutations in the genes that it encodes) would have an instant catastrophic effect, not just in one place but throughout the whole organism. If any word of the 64-word dictionary changed its meaning, so that it specified a different amino acid, just about every protein in the body would instantly change, probably in many places along its length. Unlike an ordinary mutation, which might say slightly lengthen a leg, shorten a wing or darken an eye, a change in the genetic code would change everything at once all over the body, and that would spell disaster."[12]

In this light the question becomes where did this immutable code come from and what has caused it to remain unchanged over billions of years? Why doesn't it mutate and change over time, when the living genes it encodes do exactly that over and over again? That constant DNA code combined with mutational gene change has produced the endless evolutionary advances that have taken biological life from that first single living cell to the now vast array of life that after billions of years are its offspring.

Thus, evolution is a function of change. Using its DNA code, that very first cell replicated or copied itself in order to produce a second twin cell and a third followed by a fourth with each of those cells doing the same thing over and over again. This eventually led over a vast expanse of time to oceans of algae and marine life as well as the earliest land plants.

With each generation of replication and reproduction, the existing lines of biological life tried to copy themselves using the DNA blueprints they had inherited. But every now and then there would be a flaw in this duplication process. An error would occur we think of as a mutation. Sometimes the mutation would be negative in its impact causing the end product to be less able to survive or replicate itself. These negative lines would then tend to die out. Other times the mutation would be positive in its impact and make whatever the life form was more adaptable, more capable of surviving and reproducing itself. Over the millennia this process produced creatures, whatever they were, that got bigger, stronger, faster, or in some other way more adaptable and able to survive in their environment, yet all the while their underlying DNA code system remained exactly the same as the one found in that very first cell.

So for the last 3.5 billion years this incredible system of positive mutation, based on just one DNA code, has thrown off endless numbers of ever more complex creatures and plant life, each one the product of genetic change. All this life was also somehow melded into an interdependent living complex, which, at one and the same time, both

competed with and preyed on each other while also ensuring the overall survival of the biological whole that is life itself.

Periodically, however, a culling process intervened to weed out
this vast parade of life, keeping it from overwhelming both the planet
and the rest of creation. In the last 540 million years science has been
able to identify a series of what are known as extinction events. So as
far as we know, there were at least five major ones. These events wiped
out huge portions of the then developing lines of plant and animal
species. Interspaced between these events were a regular pattern of
die-offs in which innumerable other species rose through the processes of evolution, flourished under whatever the then existing conditions
were, followed by their decline and disappearance from the fossil
record.

One prime example of such an extinction event can be found at the
close of the Permian period some 251 million years ago. In what was
a disaster for the future of much of the life that existed at that time, it
is estimated that ninety-six percent of all marine species, seventy percent of terrestrial vertebrate species, and eighty-three percent of all
insects generally disappeared from existence.

The best known of these extinction events, however, occurred
65 million years ago at the end of the Cretaceous Age. At that point a
massive asteroid plowed into Earth in what is now the Gulf of Mexico.
The resultant gigantic impact explosion blasted incredible amounts of
dust and debris into the atmosphere. Additionally the strike's powerful
hammer blow generated planet-wide seismic shock waves that led to
severe volcanic activity over wide areas, as earthquakes pulsed through
the Earth's crust. These in turn ejected vast clouds of ash and gases
into the skies, all of which added billions of tons of nearly endless
sky-darkening pollution to this planet-wide catastrophe.[13]

In combination, all this produced a blanket of gases, dust, and ash
plus impact debris that blocked out much of the Sun's life-giving rays
over the entire planet for years thereafter. Of course this lead to such
frigid temperatures over such a prolonged period that most of the

Earth's vegetation died away, and the biological food chain was totally disrupted. The end result was that the vast majority of the reptilian species including the dinosaurs, which until then had dominated the planet, in turn died off and went extinct in the geological equivalent of an eye blink.

But the devastation wasn't total; instead, it was just right. While killing off the dinosaurs, it allowed remnants of life to still cling to existence. And among those remnants was one small mammalian species about the size of a squirrel or rat. That small creature had been hanging on to existence for possibly the preceding 135 million years. In the face of reptilian carnivores that ate everything in sight, it had probably done so by living much of its life in burrows and rock crevasses. It is guessed that this creature was able to survive these protracted winter-like conditions that destroyed so many of the larger predators by living on stored nuts, roots, and subterranean insect life and perhaps hibernating.

So when Earth's atmosphere was finally able to cleanse itself and the world climate reverted to its more normal patterns, that little creature emerged into the light and began its own millions of years evolutionary climb to dominance. As it did so, its mutating offspring filled all the niches left vacant by the sudden demise of almost all the dinosaurs and other reptiles. It is from this little animal that most of the rest of the mammalian species that have existed in the last 65 million years spring, including all those we see around us today. And yes, this also includes us.

Thus, at every critical juncture from the beginning of time and at every level starting from the subatomic on up through the immensity of our universe, every event of positive change had to be just right to facilitate the necessary conditions for existence as we know it. Over a span of 13.8 billion years each step of advancement had to be precisely calibrated and sequenced to lead to us, humanity. If any one thing had gone wrong in terms of either the timing, physics, chemistry, or biology of the incredibly perfect chain of being that has been necessary to create our universe and everything in it – if any one link in that chain had been broken – life wouldn't exist nor would our world be as it is today. Why is that? Could it really have all been the result of an endless series of fortuitous accidents, or was some guiding force behind it all?

Chapter Seven

WHY?

The question thus becomes how did this perfect sequence of nearly endless numbers of absolutely necessary links stretching back to the beginning of time come to be? As we've already noted, there are only two possible answers to this question. Either there was and is a creator who lovingly orchestrated the universe or, as the renowned theoretical physicist Stephen Hawking and his co-author, Leonard Mlodinow, also a physicist, suggest in their recent book, *The Grand Design*, "…our universe is the product of spontaneous 'creation' without reference to any creator." (pg. 180) These two physicists, along with many others, contend that the universe popped into existence because of a series of natural phenomena which they postulate function in accordance with M-theory. It is "a fundamental [but so far unproven, and incomplete, though intriguing] theory of physics that is a candidate for the theory of everything."[14]

Under this theory our universe is, in effect, the product of serendipity, happenstance, or just an accident. If M-theory is correct there is the potential for a multiverse consisting of a vast number of universes that all blasted into existence at the same time. It would have only taken one, out of such an endless number of universes, to just by chance get everything right out of all those other parallel universes to produce us. We just happen to be in the exactly right one to experience it, but according to Hawking and his compatriots, given the theoretical possibility of such improbable, but natural causes for our universe, that universe cannot possibly be the product of a sentient force or God, a God they very much doubt exists.

Although Hawking would have probably argued to the contrary, it would seem equally as likely that all this perfection, piled step after perfect step, one on top of the other in such an incredible sequence, is the product of otherworldly intelligence and power which can only be denominated as God. The probability of it being the result of an equally long sequence of random chances all lining up in exactly the right order seems so vanishingly small as to be an impossibility. Even the most adamant advocates of our creation as a function of happenstance and serendipity, would, in light of all these facts, you would hope have to admit at least the possibility of divine intervention in all this.

Many of those who want to deny even the chance of there being a God seem to base their beliefs on the concept that if God actually existed we twenty-first-century humans with all our scientific and intellectual advances should now be able to physically detect, quantify, and measure that existence. But even as they do so, they have to acknowledge that even with our very best science and scientific instruments, all the matter and energy we can directly detect today constitutes less than five percent of the total energy and matter that must be present in our universe for it to exist and function as science has demonstrated it does.

Modern astrophysicists and cosmologists have come to the startling conclusion that there are incredible, but directly undetectable, forces and gigantic amounts of unseen matter at play in our universe that dominate its history, development, and future. They refer to these as "dark energy" and "dark matter." Modern science can only indirectly detect and measure them because of their aggregate effects on the visible universe we can measure. The celestial bodies we can see behave in a manner that can only be explained by the existence of these "dark" entities. We know that they have to be there, even though we don't know what they are, because our universe would not function as it does without them.

Scientists came up with the phrase "dark matter" towards the end of the twentieth century when they were finally able to measure the speeds at which galaxies, such as our Milky Way, rotated about their central axis. When they did, they came to the disturbing realization that every galaxy in the universe was revolving at such incredibly high rates of speed that they ought to fragment and fly apart. There simply wasn't enough visible or detectable matter in the forms of black holes, suns, planets, moons, and stellar dust and debris in each of those galaxies to generate gravitational pulls sufficiently strong enough to hold them together as they rotated. Since we can clearly see that the galaxies are held together, something else had to be at work. Their only answer to this puzzle was that the universe has to contain far more matter than we humans can see or detect with any of our measuring tools. And this stuff we can't detect is ninety-six percent of the total mass of matter that's out there. For want of a better term, scientists have now come to call all this invisible stuff they can't detect, and can't identify "dark matter."

Similarly, the existence of dark energy recently came to light when astronomers determined that the expansion of our universe was not, as they had expected, slowing down due to the gravitational pull of the vast amounts of visible matter we can detect in the universe; instead, that expansion is actually accelerating. And if that wasn't stunning enough, they then determined that bodies in the far reaches of outer space are actually moving away from each other at ever increasing rates of acceleration. Something was stepping on the gas pedal, so to speak. We don't know what this force is or how it works in terms of acting on the visible matter we can see, but we do know that it has to exist. So following the nomenclature of dark matter, whatever it is was deemed "dark energy."

And it turns out that the total amount of this dark energy had to likewise, be just right. Had the amount of it that is present in the universe been greater than it is, it would have so dominated the matter and radiation we can detect that our universe's structure would not have been able to form as it has. And without the structure as we now see it, as exemplified by our Earth and its place in our solar system, life would not have come into existence.

In the same manner, we are now coming to realize that dark energy's companion, dark matter, was also crucial to forming and sustaining all the universe's galactic structures. It constitutes a sort of invisible skeleton or latticework upon which the visible-matter-based galactic constellations and formations we can detect all across the universe are constructed and strung, and the gravitational forces it generates are probably what pulled together the masses of atoms that ultimately resolved themselves into all the universe's individual suns and planets.

Together dark energy and dark matter make up about ninety-five percent of all the matter and energy in the universe. Or put another way, we humans with all our science and intellect can only directly detect, measure, and quantify five percent of what makes up our universe. As to the remaining ninety-five percent, we can only say it has to be there because without it we wouldn't exist, nor would our universe function as it does.

It seems logical that if we can't directly detect ninety-five percent of our entire universe why should anyone be surprised when they cannot directly detect, measure, and quantify God? They can, however, indirectly detect, and measure the impact God has on the universe and on us. And at its heart, that godly impact has been one focused on a set of immutable factors overlaid by an endless chain of positive change with both based on love. A love grand enough to bring into existence a universe that at one and the same time supports, nurtures, and allows imperfect biological life to grow and evolve into both us and all the rest of the bio-verse that surrounds and sustains us. Thus, one is drawn to the thought that God is a God of love who has gone to enormous effort to bring all this into existence through the mechanisms of real world laws and real world change, and we can likewise see that God has done all this with painstaking and loving patience over the vast periods of time science now tells us it took to get here from that initial Big Bang.

More importantly, all this interlocking perfection suggests that it couldn't have all just happened by chance; instead, the very strong

47

implication is that there had to be an intentional force behind it, a supreme being – God! So based on this otherwise improbable chain of events, there is a powerful circumstantial case to be made that God is real and does exist.

Chapter Eight

GOD'S GARDEN

This, however, leads us to the next question. What is God doing with all this earthly biological life that has been so patiently shepherded forward over billions of years of advancement? We can now see that God has so far been content to utilize the processes of evolution and species culling over billions of years of time to create a broad garden of life to sculpt as God saw fit.

Not unlike a good human gardener or livestock breeder, God has weeded out of God's garden and culled out of God's animal stocks those lines and species that either hindered or no longer advanced God's plan for Earth's biological life. At the same time God also ensured that the lines God wished to advance had the ecological environment they needed to both challenge them to evolve and sustain them in their evolutionary growth in order for each to achieve the potential God foresaw for them.

It would appear that the tools God has chosen to use to do God's culling and weeding were climate change, predation, massive volcanic activity, and tectonic disruption of the landscape – plus the occasional cataclysmic event such as one or more gigantic meteor strikes. All of these tools were generated through God's utilization of processes found in the natural world and universe that God had previously created using the laws of physics, chemistry, biology, and mathematics. But they were all controlled and directed by God in terms of timing, location, and magnitude.

At the same time God urged on endless cycles of biological reproduction and evolutionary competition that forced all this historic mass

49

of life to evolve into ever more complex organisms and creatures. The fundamental prod God used in this effort was each life form's driving need for an energy supply to both sustain its life and allow it to reproduce itself.

At the earliest historic stages of this life-evolving operation that fuel supply probably came from minerals suspended in the ocean habitats of those life forms or was found in the rocks to which they clung. These minerals would have been liberated into their aquatic environments by the extreme heat and pressures present under those primordial conditions. Today, we still find an astonishing amount of marine life clustered around submerged volcanic vents, known as black smokers. They are located thousands of feet beneath such depths of ocean waters that everything is in total darkness. The temperatures and pressures in which this life exists are enormously greater than any that life on the surface layers of our planet could ever tolerate, and sunlight plays no part in their existence.

In a similar fashion Earth's earliest life forms evolved in equally hostile environments until they reached the point where, after a billion years or so, they finally began using photosynthesis to generate energy from sunlight. And with that event came the resultant release of free oxygen into the atmosphere as a waste product of the life cycles of these ever growing colonies of algae, the then highest level of primordial existence.

As life continued to evolve from such beginnings into its separate streams of fungi, plants, and animals, photosynthesis remained the mechanism used by nearly everything within the realm of the plant world to generate the chemical energy necessary for their life cycles. The animal world went in a different direction, however. In order for these creatures to satisfy their energy needs, almost all of the aquatic and terrestrial strands diverged down the paths of eat or die, eat or be eaten. They followed a pattern of gaining their food energy by dining on either vegetable plant life in all its myriad forms or eating other creatures including the plant eaters. And this, as we will later see, was to have huge implications for humanity as it ultimately emerged out of the chain of animal being.

In the meantime, that eat or be eaten response produced what Darwin referred to as the "war of nature," an "ever escalating arms race between predator and prey, parasites and hosts." This competition forced mutation after mutation in the ongoing race for survival of all these creatures. But out of all this conflict there has emerged a finely calibrated equation in which green plant species outnumbered animal life forms by at least ten to one. Were this not so there would not have been a sufficient mass of biological plant life to sustain the animal world's food chain including us at its top. Thus, there had to be far more activity, diversity, and mutation down the plant lines of evolutionary growth than there were in those of the animal kingdom. So why was that? Where did this finely calibrated balance come from? Could it not be God, the loving gardener!

And God as that gardener made choices. The majority of the larger air-breathing animal lines that emerged out of this evolutionary stream were quadrupeds that, in one way or another utilized all four of their limbs for locomotion. One particular line went in a different direction, however. It became more or less bipedal and led to the primates to which our hominoid line and we humans belong. This line evolved its limbs into appendages that were capable of grasping, thus allowing them to climb as well as walk.

This was the line that God singled out for particular attention. Under God's focus, what biologists denominate as natural selection became, in effect, artificial selection. It would seem that God selected this line because it had the potential to evolve into fully bipedal, big-brained creatures, who had the capacity to become both self-aware and reasoning beings, able to manually manipulate the environments in which they existed; likewise, given its critical lack of fangs, claws, size, or speed, this line would also have to fall back on its intelligence and band together to survive in the greater eat or be eaten world surrounding it.

51

God apparently prized the primate line because its forelimbs in conjunction with its intellectual potential made it capable of both grasping objects and fine dexterity in the manipulation of those objects. It also had the potential to evolve to the point that its individual members (in the case of *Homo sapiens*) would be able to verbally communicate with each other in abstract terms and ideas that could encompass stories including the possibility of God. Using these characteristics allowed this line to progress to both tool making and ultimately the creation of cultures and civilizations. Having made this selection out of all the available choices presented among the evolving lines of biological life, God proceeded to facilitate its evolutionary growth in ways not lavished on the rest of Earth's life forms.

As the primate line was developing, about ten million years ago another evolutionary branching occurred that has led to the great apes, including gorillas, chimpanzees, and hominoids. In relation to these three primate branches, the DNA makeup of humans differs only by one and a half percent from that of chimps and a bare two percent from gorillas, making both gorillas and chimps our very near cousins.

Gorillas became the first of the great apes to peel away from this ancestral primate predecessor line. They were followed by hominoids and then by chimpanzees about two to three million years ago. In the ensuing millions of years right up until today both the gorilla and chimp lines have remained more or less evolutionarily unchanged, with only the chimpanzee line throwing off a significant mutational variant in the form of a near chimp, the bonobo, about one to three million years in the past. Otherwise, gorillas, chimps, and bonobos are essentially the same today as their ancestors were when they first appeared on the evolutionary scene despite Darwin's natural selection and its best efforts to force their further mutation.

By contrast the hominoids, our ancestors, made their first appearances in our evolutionary history perhaps six million years ago. Since that time our hominoid stream has seen at least eighteen to twenty or more evolutionary leaps that we can currently identify.

A number of these turned out to have been what are currently considered dead ends. These apparent false starts include *Australopithecus garhi* (3 to 2 million years ago), *Paranthropus robustus* (1.8 to 1.4 million years ago), *Homo antecessor* (1.2 to 0.8 million years ago), and *Homo erectus* (if it is considered to be outside our direct line of ancestry which many experts consider it not to be, 1 million to 50 thousand years ago), plus a number of others.

The twelve or so leaps down the hominoid line that are often considered to have led to *Homo sapiens* include:

- *Ardipithecus kadabba* – 5.8 to 5.2 million years ago
- *Ardipithecus ramidus* – 4.4 million years ago
- *Australopithecus anamensis* – 4.2 to 3.3 million years ago
- *Australopithecus afarensis* – 4.1 to 2.8 million years ago
- *Australopithecus africanus* – 3.3 to 2.8 million years ago
- *Kenyanthropus platyops* – 3.5 to 3.2 million years ago
- *Homo habilis* – 2.2 to 1.5 million years ago
- *Homo rudolfensis* – 1.9 million years ago
- *Homo ergaster* – 1.8 to 0.6 million years ago
- *Homo heidelbergensis* – 0.6 to 0.25 million years ago
- *Homo neanderthalensis* – 0.36 to 30 thousand years ago
- *Homo sapiens* – 150 to 200 thousand years ago to the present

It was originally thought that the Neanderthals were also a dead end, but recent DNA evidence extracted from Neanderthal fossil bones indicates that they contributed between one and four percent of the genes of the *Homo sapiens* populations living outside of Africa.[15]

Most recently a new discovery in South Africa of *Australopithecus sediba* (2.0 to 1.9 million ago), which had both ape-like and hominid characteristics, has given rise to a hotly debated discussion as to whether this creature represents the beginnings of the transition from ape to human.[16]

Each of these predecessor species made its appearance on the evolutionary stage, played its part, and then disappeared, leaving no primate competition for us. All the while they and our near cousins, the gorillas and chimpanzees, hung around this entire timespan of millions of years with little or no change.

Even though anthropologists and biologists will argue endlessly over whether one or more of these species were really separate and apart from some of the others, or in or out of our ancestral line, we do know that *Homo sapiens* sprang out of this chain approximately two-hundred thousand or so years ago. As the new kids on the block, so to speak, we are the result of all the preceding eleven or so leaps in evolutionary change that had moved forward along what were apparently rather messy sequential and sometimes parallel or overlapping predecessor lines, before merging back into the main stream of our evolutionary heritage.

While only biologists and anthropologists could love these convoluted and esoteric names, one constant in this stream of life seems to be the progressive growth in both body and brain size from one species to next in this our line of evolution. So we have to ask, why were there these extraordinarily frequent positive leaps of evolutionary change in this one primate line when the parallel lines of our cousins, the chimps and gorillas, remained essentially static? Could it not, in effect, have been artificially encouraged as opposed to pure natural selection? And if it was, who or what did that artificial selecting if it was not a loving God who paid special attention to the evolutionary processes that produced us?

In terms of *Homo sapiens* it is, as we will see, important to note that as much as a quarter to a third of our adaptive evolution has occurred within the last forty thousand years while our great ape cousins continued to stand still.

As we've noted, there are many who are convinced that none of this was driven by any intelligent force. Instead, to their way of thinking, it had to be purely the product of blind non-thinking Darwinian natural selection. However, this belief ignores some of the most inter-esting parallels between the evolutionary path that led our progenitors from great ape status to *Homo sapiens* on the one hand and our own artificial human sculpting of the plant and animal worlds surrounding us on the other. This artificially created impact on other life forms produced by we humans, who are infinitely less than gods, has striking similarities to the adaptive evolutionary advances we have experi-enced as a species.

So let's explore those parallels. Using man's best friend, the dog, as one of his prime examples, Richard Dawkins in *The Greatest Show on Earth* demonstrates how we humans have lovingly turned wolves into Chihuahuas. Starting thirty to forty thousand years ago with the wolf, humans began selecting for traits we found in them that we prized. These included, for example, loyalty, the ability to herd other animals, to track by scent, or act as watchdogs. We have likewise focused on size. Human efforts have produced dogs ranging in size from the Great Dane and Irish wolfhound to the Pekinese. By selec-tively encouraging the breeding of generations of what became "designer dogs" with the particular characteristics some group of hu-mans desired, we have, in Dawkins' words, "sculpted the gene pool" of dog-dom.

As part of our doing this to man's best friend, we ignored a host of imperfections in these animals we love so much, even as we sculpted them in favor of those characteristics and traits we wanted. The fact that they live short lives only made it easier to produce generational change in less time. So who cares since we can breed more of them? Bad hips or doggy breath; no problem. Not necessarily the brightest of animals; that's OK, we still love them. We went for what we humans wanted in dogs and got it in less than thirty thousand years of our arti-ficial selection efforts. Today there are over two hundred different rec-ognized breeds of dog all distinctly differing in size, configuration, and abilities. And this figure ignores the millions and millions of just plain mutts, which are often the best of the lot. But the one thing they all have in common is their thirty-thousand-year descent from their original wolf forbearers.

The wolf, by the way, was more than complicit in this sculpting process. Starting those thirty-to forty-thousand years ago, the wolf went from wild hunter to camp scavenger to camp dog. Generation by generation the wolf, or more precisely some wolves, participated in domesticating themselves.

First they scavenged human camp refuse dumps. They thus became accustomed to being around humans, being played with by people when they were puppies and chasing after the food scraps those camp folks tossed to them. As a result of this behavior they began to forget how to hunt and, instead, became dependent on humans for their food and protection from their own wild brothers and other predators. They became our domestic partners in shaping who they were and what their future evolutionary growth would be.

We have likewise done the same thing with all sorts of other plants and animals. Just consider sheep, cattle, pigs, chickens, corn, and wheat as other prime examples of our successes at artificial selection over the same time frame. Over the millennia we humans have sculpted their various gene pools to change them from wild plants and animals into the far different domesticates humanity now depends on for sustenance and clothing, and we did so through forms of artificial selection and control that we directed.

By contrast, Dawkins and his co-believers argue that a god could not in the same way have intentionally sculpt the primate gene pool to achieve that god's ends and aims for what became *Homo sapiens.* However, the only difference between the two processes is that God was willing to take several million years longer in the process than the thirty thousand or so years we have spent morphing wild wolves into Pugs and German shepherds through artificially directed evolutionary change.

Arguably, God did the same with us! And God did so even as God ignored or put up with our faults and imperfections, which were as great as, or perhaps even greater than some of those found in the other alternative evolutionary lines of earthly life available to God for this effort, including our near cousins the gorilla and chimpanzee. As we've noted, over the same periods of time they and so much else in Earth's cornucopia of life remained relatively unchanged despite evolutionary pressures and the forces of natural selection even as our line evolved at a startlingly rapid pace.

But as proof of such a deity's nonexistence, a number of those who deny God's involvement in all this, or even God's very existence, point to the myriad of imperfections in the physical design of both *Homo sapiens* and the other life forms surrounding us. They also argue that the horrific trauma that runs rampant through so much of daily life across Earth's biological spectrum demonstrates that nonexistence. According to authors such as Dawkins and Christopher Hitchens, for example, if God existed, God would have done a much better job of creation in terms of both the design of the world's life forms and the functioning of the universe and biological life.

To their way of thinking, a perfect God would have perfectly served us and the world by having perfectly designed each and every species. For them a real God would have created an exactly engineered ecosystem. That is a system in which prey and predators would have been long-lived and equitably balanced in terms of the minimum amounts of predation, with its accompanying trauma, needed to sustain each species and the system.

Dawkins argues that if there were intelligence behind the evolutionary process, predator and prey species would have reached accommodation. For him, this accommodation would have produced a cessation of the evolutionary competition in which each life form attempts to increase its individual capacities to either kill or avoid being killed. They would as a result have ceased or vastly slowed their rates of mutational changes that fuel evolutionary growth. If a perfect God had been the author of all this, the various species, having already been perfectly designed, would give up their evolutionary arms races, so to speak, and advance at a much more sedentary pace in which each would remain in harmonious equipoise as would the entire ecological system. As a consequence there would have been far less trauma and terror associated with biological life.

Since that is clearly not the case, for them there can be no God. Instead there is only random chance in the world and natural selection, with its eat or be eaten, red of tooth and fang imperative, co-joined with the accompanying drive towards unlimited reproduction of each species. With Darwinian evolutionary competition as the proven law of nature there can be no intelligent designer who underlies this endless predator/prey-based system.

Dawkins also illustrates how, when we examine the human body, we find a whole series of flawed and imperfectly engineered designs in terms of our various organs and bodily wiring. He adds that the same holds true for the rest of the animal kingdom, providing stark examples to illustrate his thesis. Working from these examples, he contends that such flawed design work must prove the nonexistence of God because, again, a perfect God would have perfectly designed each species as well as the biological whole of earthly life.

Because a number of other prominent authorities share this position with him, let's examine his and their positions. In doing so we first have to ask would God's apparent purposes (as we've come to know them by examining God's use of the natural world and its laws) be served by perfectly designing and engineering humans, and the human body, plus all the other animal species as well as perfectly balancing the entire biological world in order to create absolute harmony.

As we've seen, God did not elect to initiate biological life by designing each plant and animal such that it was perfectly engineered and then will it into existence as part of a perfect whole; instead, God chose the violently competitive approach of natural evolution, starting with life's single cell beginnings and then forcing the various species to evolve in terms of their capabilities, intelligence, and often size.

Since God decided to utilize real world laws and natural processes on an incremental, step-by-step basis to implement the divine plan, it would not have served God's purpose of encouraging such incremental growth through mutation to have it vastly slowed down or even halted due to the absence of species competition. Replacing that competition with benign harmony and long life through perfect engineering would have hindered, not helped God efforts. Instead, God only required each creature, each plant to have a life span sufficiently long enough to fulfill the function God had for it in the climb of life from those single cell beginnings to the complexity of *Homo sapiens*, even as flawed as we are. Thus, the counter argument is that earthly biological bodies are instead designed to not last past a certain point. This ensures that each individual biological life would have to periodically be replaced after a reasonable period of existence.

These facts would indicate that what God has done is use God's version of artificial selection to force more rapid and selected forms of evolutionary change when and where it suites God's purposes. As species evolved over broad fronts of life, God searched their lines to pick out those whose evolutionary growth potential would produce a world, in our case Earth, which met God's long-term plans. At the same time God culled and pruned away those lines that would inhibit this process. The great asteroid strike and its resultant extinction event at the end of the Cretaceous period 65 million years ago, marking the demise of dinosaurian domination and the rise of mammals, is the most obvious example of that artificial selection process at work.

Down through the billions of years of evolutionary reproduction we can see that God has likewise lovingly focused on at least one line so that it would survive through die-offs, extinction events, and the predatory attacks of other species. And that line is the one that led to mammalian primates and ultimately humanity. As we've demonstrated, this line, especially when it reached the stage of the hominids, has leaped forward at a pace that vastly outstripped its surrounding species lines.

Arguably, these comparatively rapid leaps were the result of God's prodding that advance through the mechanism of artificial selection. We humans have mimicked that process in our own artificial evolution of the plants and animals we've needed to sustain our climb from savagery into civilization, as exemplified by our morphing the wild wolf into today's lap dogs.

For those who choose to think that humanity should have been perfectly designed by a perfect God in order to prove God's existence, we also must ask the question, what makes anyone think we humans are or should be the end and apex of God's efforts in the march of this artificially driven selection process? Given what we know of the billions of years old evolutionary history we can see behind us plus the fact that God has intentionally situated our Earth and us in a Goldilocks zone with a Sun that can sustain the evolutionary advance of *Homo sapiens* into *Homo futurus* and beyond for several billion more years, isn't it more likely that we are only a way station along God's planned line of advance and that our progeny has a long way yet to grow towards a perfection only a loving God can see? And finally we need to ask what makes us think that God should serve us and our purposes as opposed to humanity serving God and God's purposes?

Chapter Nine

FREE WILL AND GOD'S SILENT VOICE

This analysis of the hard facts that are available to us suggests the presence of an eternal, loving, and very patient God. A God who uses finely calibrated laws of quantum mechanics, physics, chemistry, and mathematics – plus the processes of biological evolution over the span of billions of years – to create both our universe and our Earth with its vast garden of life, and us. It's a picture that also suggests that God has a preference for these natural forces and laws to work God's will, as opposed to wielding some version of a godly magic wand to just order into existence the end results God wishes to achieve.

Yet despite this preference, it is also strongly arguable that, from time to time, God does step into the picture to exercise the directing control needed to steer events towards the outcomes God desires. With such interventions God ensures the execution of God's overall plan for all that God has created.

A very likely candidate for an example of such direct intervention is the gigantic asteroid that may have plowed into Earth sixty-five million years ago to end Earth's Cretaceous Age. This devastating event rearranged the order and hierarchy of the world's biological life forms. It was a cataclysm that fit its own "just right Goldilocks definition" of being just massive enough to wipe out the dinosaurs (and their rapacious, eat everything in sight, domination of the then existing food chain) but not so big that it totally destroyed all the rest of Earth's life. As we've noted, included among the life that managed to survive this cataclysm were the struggling mammalian lines that had previously

61

cowered in hiding out of fear of the then dominant dinosaur predators. The disappearance of their reptilian enemies allowed these mammalian lines to evolve into Earth's newly dominant species. The upshot left we humans at the top of that chain. This was an event that never would have happened had the dinosaurs survived.

Within this framework one of the principal questions that needs to be considered is, apart from possibly flinging giant asteroids at Earth, how did and does God intervene to shape life? How did God go about sculpting the individual species found in God's garden with a special focus on the primate lines that have morphed into today's humanity? The coming chapters will discuss some possible answers to this question. And the first of these suggestions we will be looking at is "God's silent voice."

Over the millennia we and our ancestors have often looked and longed for God to simply say out loud, in a way we could all hear, exactly what God wants us to do and become. All God would have to do is to audibly belt out verbal commands to us, and we'd probably salute and then march off smartly in whatever directions God ordered us to go. The end result would be certainty. There would be no question whether or not God actually existed, what we are meant to be and do, or for that matter, how to worship God.

But that hasn't happened on anything approaching a mass audience basis; instead, those who have actually heard God's voice in any sort of an audible way are few and far between. So we need to ask ourselves, why is that? Why hasn't God directly delivered instructions to us in clear verbal messages we could all understand, knowing without doubt that they came from a divine and omnipotent being?

Based on what most closely approaches a historic record, the answer to this part of the puzzle is that when God does verbalize commands to individual humans it seems to overwhelm them and their free will. Looking at just the Biblical writings of some of the Old Testament Prophets and in the New Testament letters of Paul, we find repeated confirmations of this phenomenon. The Prophet Amos emphatically states, "The lion roars – who will not be afraid? The Lord God speaks – who will not prophesy?"[17]

Jeremiah bemoans bitterly:

"You duped me O Lord, and I let myself be duped; you were too strong for me, and you triumphed.[18] All the day I am an object of laughter; everyone mocks me. Whenever I speak, I must cry out, violence and outrage is my message; The word of the Lord has brought me derision and reproach, all the day I say to myself, I will not mention him, I will speak in his name no more. But then it becomes like fire burning in my heart, imprisoned in my bones; I grow weary holding it in, I cannot endure it."[19]

In this same vein, Paul emphatically declared to the Corinthians that, "I am under compulsion and I have no choice. I am ruined if I do not preach it"[20] referring to the Gospel of the Lord.

So clearly then, the people who actually hear God's voice in an audible or near audible sense comply with its commands, whether they want to or not. It apparently exercises a sort of pressure on them that they cannot ignore or escape. In effect their human free will is compromised, if not completely destroyed. And from God's stand-point this apparently is not something to be desired on a mass basis across all of humanity.

Why? Because were God to use this overwhelming power of command to audibly compel our thoughts and actions, humanity would be no more than a world of puppets dangling from God's verbal strings. We would be forever involuntarily dancing to God's controlling tune. Like Jeremiah, God's audible words would burn in us, and no matter how hard we tried, ultimately we could not resist. We would be under compulsion just as Paul was.

If that were the case, then from its first evolutionary glimmerings, humanity would have spent life waiting expectantly for the next command from on high. And that would have absolutely stunted human intellectual and moral evolution. Were we dependent on God's audible instructions, we would never have truly understood why it was important to move out of our ancestors' animalistic state of not recognizing the consequences and possibilities of their conduct. We'd never have approached the point where, using our free will, we can now understand what evil is and how our conduct can be either good or bad. Humanity would never have begun the process of working its way through those flaws which arise out of our biological heritage. That heritage is one of eat or be eaten and doing whatever we had to do to survive and procreate – no matter the consequences to the rest of the life in God's garden; instead, based on godly compulsion, we'd have simply done what we were told without understanding its real importance or consequence.

So it would seem that rather than taking the easy road of direct command, God has chosen the far longer path of evolution in order to foster the growth of human free will. It would likewise appear that the reason for this divine election is that free will is something that God prizes in humans. Why? Because with free will we can choose to do what is right or, alternatively, to follow the path of wrong based on our own individual and collective decisions as sentient beings. We can choose to believe in God or not, as we find the light to do so within ourselves. And this, it would seem, is something a loving God wants us to come to on our own, free of godly compulsion.

At the beginning of this trek, protohumans were more or less like all the other animals that surrounded them. At best they moved about looking for their next meal or the chance to mate. The strong dominated the weak without reference to right or wrong.

But as hominids evolved into *Homo sapiens* we began to develop a conscious free will that could and sometimes would counter the animal instincts that had previously governed our lives. We started to recognize, in consciously thought out ways, that there were alternative paths that advanced not only the acting individuals, but also the rest of life surrounding them. Our ancestors came to realize they could choose to protect the weak, the old, and the lame. They could elect to not fight their own kind for food or territory but, instead, choose to cooperate with them in obtaining those desirable things. They started to differentiate between what constituted good conduct and what was detrimental. And they could choose what to believe in terms of what they saw and experienced as forces outside of the real world they directly encountered on a day-to-day basis.

It was this capacity for free will analysis and decision making that has been crucial to human growth and survival from our earliest hominid ancestors to what we are today. Use of their intellects in combination with their capacity for free will allowed humans to adapt to shifting food supplies and geographic locations when climate change, overhunting, or plant and animal disease threatened to obliterate their accustomed dietary sources and preferences or other disasters menaced their existence. So this capacity has been critical to both human survival as well as our spread across most of the world.

Since it would seem God has gone to great lengths to foster our free will as one of the major aspects and attributes of human evolution, it would appear that it is something God prizes. At the same time, instead of simply taking a hands-off attitude that let our evolving conduct follow whatever course chance and our free will pushed us towards, God has, out of love, chosen to mold us for good. But to do so God has had to find ways other than direct command to both influence and teach us what good and right are without suppressing our capacity for such free will thinking and conduct.

Assuming God wished to influence humanity's overall conduct and thinking in positive ways other than by direct command on a mass basis, we should look at what are the more subtle alternatives available for a loving God's use. The first of these is "God's silent voice." Utilizing this silent voice, God is able to communicate with us at the deep sub-vocal level of our subconscious. It would seem God is able to access that part of our minds, even though we do not consciously perceive such communications as direct thought.

65

At this level, that silent voice acts more as a suggestion than a command. As such it is not all powerful or overwhelming, but it does exert pressure and direction on our personal conduct even though we can still ignore or disregard its suggestions based on the exercise of our free will.

Obviously, not all or even a majority of the silent messages, impulses, unbidden thoughts and ideas, whims, or answers to questions that jump into our minds seemingly out of nowhere are the product of God's silent voice. Rather, from time to time, when God wants to influence or teach us, God can insert this silent voice into what is a stream of communication constantly welling up out of our psyches into our conscious thoughts. It is up to us to then sort out the differences between the two and chose to act or not act on those coming from God.

But once we've unconsciously heard that silent voice, its message percolates up into the conscious levels of our mind as a thought or suggestion out of nowhere. Think about it. How often has each of us responded to silent urges, impulses, or ideas that seem to pop unbidden into our heads? Each is a sort of silent voice calling for action or perhaps reflection. We are moved to give to a beggar, help an animal in distress, to look around a corner or out a window for no reason other than what we think of as a whim, only to see something that matters to us. We have all spoken to someone who attracts us without saying a word; likewise, we will pick up a book thinking of it as a random act or read an article assuming doing so is pure chance only to find those encounters have an impact on our lives and thinking. We may elect to turn down a side street for no particular reason and then see something unexpected that affects us or run into someone with whom we have a surprising, but meaningful, conversation. And many of these events lead to thoughts, concepts, or actions that influence our lives and the lives of others.

How many times have we or someone we know experienced what is later thought of as a "calling." In its grip individuals experience an ongoing urge that is ultimately very hard to ignore. The called individuals may be drawn to a profession or vocation: say to become a doctor, a teacher, priest, minister, or nun; to serve in the military; or to commit themselves to some other form of public service. In similar ways people can feel drawn towards a particular town or geographic region to settle in simply because this is where something tells them they should go. In a similar fashion, they may be attracted to a particular faith that suddenly feels right to them even if it is totally outside of what they grew up with.

And finally how many of us have felt the impact of what we think of as our "conscience" silently insisting that something we've done or are contemplating doing is wrong, or that same conscience nags at us about doing the right thing when the right thing is something we really don't want to do or are afraid of, but know we should do. What better way for a loving God to speak to us without speaking?

In response to all this we are inclined to put our actions and thoughts down to impulse, whims, urges, or instincts. We can even act on them without thinking about why we are doing what we are doing. We just feel a need to do so that can often be more powerful and compelling than a lot of the decisions we arrive at through the processes of conscious thought.

Knowing this, people shouldn't categorically deny that God could use these mechanisms as silent vehicles to push us towards thought or action, to draw us to beliefs or moral codes, or to lead us to ideas. What more logical way is there for a loving God to insert directions and desires into our lives in ways that still allow us to exercise our free will to either act or refrain from acting on God's suggestions?

This silent voice would also be a powerful tool in God's sculpting of the human gene pool and our evolution. Through it God can steer individuals into situations and relationships that facilitate God's long-term plan. For example, each of us can probably think of the paring of individuals who seem totally unsuited to either living or working together, but their union or partnership has produced offspring or re-sults that go on to benefit the rest of humanity despite the incompati-bility of their personalities. Thus, we should at least consider the possibility that God has silently pushed them together as part of shap-ing the human gene pool or condition in ways that advance God's plan for us.

Chapter Ten

GOD'S CHALLENGES

Given humanity's stubborn nature, God's silent voice alone would not seem to be nearly enough to work all the wonders God has achieved in advancing our growth as a species. So what are some of the other mechanisms God has used to produce these wondrous results beyond those we've already identified? And once we identify them, we also need to ask why and how has God used them?

When you look at the history of evolution and of humanity, as well as our own lives, one strategy that should instantly command our attention is God's constant and consistent use of physical, intellectual, moral, and spiritual challenges throughout both evolution and human development. Setting aside, for the moment, the fact that God is the ultimate challenge for each of us – a challenge posed in every sense of the word – we need to come to understand why God has so heavily utilized challenges as a means of inducing advancement in both the evolution of life and human growth.

Implicit in everything we've looked at to this point is the fact that evolution itself is a process of challenge-induced change. As we've seen, every life form that has ever existed has been challenged on a daily basis to find the energy sources it has needed to sustain life and allow it to procreate. The struggle to meet these two overarching challenges has forced each and every species to engage in the process of mutational change, as God has set ever higher the bar that every generation of life has had to surmount in order to survive.

Consider that the Darwinian war of eat or be eaten is the product of this God-ordained challenge. Each prey species has struggled to

69

find new ways to defend itself through mutational change. Through the processes of evolution each species seeks to sharpen its senses, increase its speed of flight, thicken its hide, or get higher off the ground and hide better through camouflage or habitat. In the face of these challenges predators have likewise been challenged to defeat these advances with their own parallel evolutionary changes.

One of the major evolutionary advances that accompanied this progression of change has been the increase in brain size in a significant number of species spread across the entire spectrum of biological life forms. This is especially true in terms of the primate lines in general and the hominids in particular. This increase in brain size and presumably intelligence became the major force behind our growing human ability to cope with a never-ending array of challenges as they've been presented to us by God.

Even as hominids (as both predator and prey) were using their brains to meet the challenges posed by the other species God surrounded them with, they also had to use their intellects to deal with the threatening challenges found in the natural world God had structured for their habitat. Looking at that world it becomes apparent that the challenges it presents run in a sort of just right Goldilocks spectrum of threat levels. These threats include fire, flood, famine, pestilence, plague, drought, extreme winter weather and summer heat, storms, tsunamis, tornadoes, hurricanes, and earthquakes plus a host of others such as insect infestations and plant disease that devastate food supplies.

At first glance all these menaces can seem to be overwhelming; however, a more discerning analysis brings the realization that, over the time frames of evolution, the severities of these threats have been held within parameters that were sufficient to force our human ancestors and us to spread out across almost the entire face of the Earth in our efforts to find more hospitable settings in which to live and grow. At the same time these challenges have not been so great that they could not be overcome by sufficient numbers of our forbearers to ensure the survival and continued upward evolution of our species. They were, in short, just right in terms of forcing us to grow while at the same time not being so devastating as to crush the potential for that growth.

A human analogy for this balance can be found in physical fitness trainers. A good trainer utilizes ever more rigorous and changing routines and obstacles to force those they are training to build up and expand their physical abilities, stamina, coordination, and muscles. While doing so the best trainers are careful to avoid injuries to their trainees that would permanently halt or reverse the growth the trainer had been fostering, but soreness and discomfort during the training regimen are perfectly OK.

This is a good metaphor for what God is doing with us in terms of the finely and lovingly balanced environment of layered challenges God has confronted us and our ancestors with, even if we don't recognize it as such. God has used these challenges to ensure we do not stagnate, to instead guarantee that we continue to evolve toward the outcomes God wishes us to achieve.

In doing this, God is not being sadistic, rather, God loves us enough to ensure we do more than merely exist like a herd of placid, but flawed, cattle who spend eternity, heads down, grazing on meaningless and unfulfilling fodder that does nothing to force us to think, feel, cope, and expand our intellectual, moral, and spiritual horizons. God instead cares about us enough to allow us to experience all of life in its full range of excitement, rigor, and challenges as well as its pain, boring repetition, and drudgery. This necessarily means that we have to deal with all the good and bad that make those lives complete, and in doing so we grow individually and collectively as we chose our responses to each obstacle and challenge we encounter.

Perhaps the most important form of challenge God has confronted us with in order to work us out of our imperfections and the baser aspects of human nature is the sheer diversity of humanity itself. At the forefront of this set of human challenges are those presented to us by our gender differences. These, by the way, are not just limited to challenges created by the differences between male and female. They also include the accompanying contradictions inherent in our various sexual orientations within each gender, which set up their own host of interpersonal human conflicts. The God who loves us confronts us with all of these challenges because God wants us to learn to love one another, no matter what our differences are, in the same way God loves each of us.

Next are the difficulties generated by the necessity of our living together and sharing in families, groups, tribes, villages, cities, regions, and nations in order to both survive and flourish. All these shades of interaction come with their own accompanying levels of challenge: the requirement for cooperation, acting with toleration, and overcoming conflict, and competition for finite food supplies and other resources. We must continually face all of these while trying to raise our progeny from helpless infants to fully functioning and procreating adults who can interact in loving ways. And to all these realms of *Homo sapiens* interaction we must also add the issues of clashing and conflicting ideologies, faiths, politics, and systems of governance.

Without challenges, especially in their human relationships, people develop unwarranted certainty in their thinking and beliefs. This leads to a dangerous rigidity in who and what they are – a rigidity that produces deformed and negative growth away from God's intended path for them.

This is why we need minorities in all our societies and schools of thought – minorities that challenge us to reexamine our received certainties of belief and conduct. Absent such minority challenges, people are much less likely to periodically reevaluate who and what they are – what their relationships with each other should be. Without that reexamination, schools of thought, societies, and individuals are prone to become petrified caricatures of what God's plan for them is.

As we look around the world today, it becomes apparent that excessive homogeneity and conformity of thought stultifies both human growth and the positive intellectual activity that produces new ideas. When humans seek to impose such uniformity in group composition, thought, and belief they end up with the Talibans and ISISs of the world.

God recognizes this and it is one of the reasons God seems to foster diversity in all aspects of life including humanity. It would also seem that because of this God doesn't want us to just accept the way things are; instead, God encourages us to challenge what we think, believe, and do to see if those thoughts, beliefs, and actions will withstand those challenges, or whether they need to be modified and updated.

Inherent in all these challenges arising out of our human conduct and condition is our need to recognize, confront, and overcome evil and the baser side of human nature (both of which we will discuss more fully in the next chapter). God has spent tens of thousands of years patiently going through the processes of teaching us what evil is, how to recognize it, and how to deal with it – not only in ourselves but also in the rest of humanity both near and far from us.

Apparently our reward for mastering all these interlocking layers of challenge is the opportunity to physically, intellectually, morally, and spiritually confront even more challenges. So we have to ask the "what" question again. What is God accomplishing by pushing each of us through these lifelong obstacle courses?

The most obvious answer is that God wishes each of us to grow individually, collectively, and generationally. We grow in our thinking and intellect. We grow in moral and spiritual understanding. We grow in our abilities to live with each other – as individuals and as societies – in ways that are mutually supportive as opposed to destructive. We grow in understanding and toleration. We grow in our focus on both loving and protecting all life, not just human life.

The less obvious answer, however, is that we are also forced to encounter as real experiences the dark sides of all these challenges. We have to actually deal with the real events of adversity, want, hunger, insecurity, intolerance, hatred, sickness, disease, greed, conflict, pain, fear, danger, death, apathy, change, and a host of other debilitating conditions. The negative sides of all of these are threats to our ultimate growth that we must first recognize and then face and overcome.

Let us also not forget that ease, comfort, complacency, wealth, safety, and satisfaction in all its forms – especially self-satisfaction – present their own challenges that, in the long run, are equally as difficult for us to master. In short, one way or another, all of human life is a challenge no matter which aspects of it each of us are faced with.

If people have any hope of achieving a real relationship with God, they will have to come to grips with the experiential actuality of some combination of all of these aspects of our natures and existence, as opposed to only having a theoretical or academic acquaintance with their possibility. This is because the part of God we wish to connect with understands, on both the emotional and intellectual levels, the reality of each of these through God's own ultimate connection with all life and all that life has ever faced or suffered. Absent the living knowledge of all these challenges we can never hope to make the right choices as opposed to the easy or comfortable ones when God asks us to utilize our free will to do what God needs us to do in whatever ultimate future God has planned for humanity both collectively and individually. Without these experiences we will never have the capacity to meet God's expectations of us. We will never be able to fully engage with God.

A gross analogy to this experiential problem can be found in the consequences of the human decisions leading to and following the acts of dropping the two atomic bombs at Hiroshima and Nagasaki that ended World War II. Setting aside whether the event of those bombings was right or wrong at the time, without those decisions and their hideous aftermath, humanity would have had no concrete real world experiences of the true horrors of nuclear warfare to inform their decision making in an age of potential global atomic conflict. Our leaders on both sides of the subsequent Cold War, with its precariously stalemated nuclear arms race, would have had nothing more than an academic understanding of the devastatingly destructive impact on the entire world of an all-out thermonuclear exchange between the two sides.

In the minds of many, hydrogen and atomic bombs would have been just bigger versions of the bombings of WWII. And either or both sides could well have been less deterred, or not deterred at all, in terms of pulling the nuclear trigger in confrontations such as the Cuban Missile Crisis of 1962. Only the actual horrendous destruction and suffering the world had to confront face to face in the radioactive rubble of Nagasaki and Hiroshima provided the examples needed to keep another nuclear holocaust from happening. Apparently, we will need something akin to this sort of real understanding of all the downsides of the various experiences we have been discussing in order to cope with the challenges the potential of a perfect eternal life with God will present to us. In that eternity these experiences may keep us from looking back nostalgically at our former earthly lives as we try to handle whatever challenges still await us in terms of an eternal existence in the presence of God.

Thus, given the march of evolutionary life and our own *Homo sapiens* history of never ending challenges, we can only assume that it would not have served God's purposes to have made life and us perfect from the start. Instead, we can surmise that for reasons that advance God's plans for us, God requires us to have had and to continue to have all these real world challenges and learning opportunities that we and our forbearers have had to meet through the exercise of our free will, our wits, and our experiential knowledge. The actuality of all these challenges has forced us to grow and evolve in ways that a merely theoretical or academic understanding of them never could have. All of this has real implications for our efforts to try to start to fathom what we can begin to see is, at bottom, a loving, but incredibly complex God who cares about all of humanity for which God has a purpose and plan.

Individually and collectively, however, our relationship with that loving God is our ultimate challenge.

Chapter Eleven

OVERCOMING EVIL

Another major challenge God uses to advance the growth of our *Homo sapiens* species is our God-induced capacity to recognize and confront evil on both the intellectual and emotional levels. To understand God's use of evil as an obstacle we have to push against in order to grow – as God would have us grow – we have to first define what evil is and is not, as well as where it comes from. Evil itself only exists when and where there are living beings that have the capacity to comprehend its existence. As far as we know, here on Earth, we humans are the only biological creatures who can do this.

Without such recognition evil is not an issue; instead, the facts of the natural world, with all its dangers, and those of biological life, in the "eat or be eaten" sense, are just that, facts. Fire, floods, droughts, hurricanes, earthquakes, blizzards, tornadoes, and the rest of the pantheon of natural disasters are unthinking events that have no capacity to be evil even though they may be incredibly destructive, terrifying, and life threatening. As a result we may want to demonize such menaces, but that does not make them evil.

In terms of living conduct, with the exception of we humans, the biological world's predators have no concept of the moral questions posed by the suffering experienced by the prey they are devouring. Those predators are just responding to the driving compulsion hardwired into their DNA to satisfy their biologically induced hunger needs for the food energy provided by the flesh of their prey. In doing so they are not engaged in acts of evil. They are only meeting the demands evolution has imposed on them to survive and perpetuate their

77

species. Using the only tools evolution has given them to satisfy those needs, they have no choice other than to employ cunning, claws, fangs, teeth, speed, and muscle power to rend, tear, and devour whatever food sources are available to them. Their only alternative is slow death by starvation.

In turn, their victims can only respond with counter speed and their own cunning, plus horns, hooves, counter strength, and numbers in their efforts to overcome predatory attacks. But without the depredations of their adversaries, the component numbers of the individual prey species would so multiply that they would obliterate the vegetable food supplies our Earth can produce to sustain them. This would obviously lead to their deaths by starvation. None of this, however, is the subject of any sort of moral analysis of right or wrong by either predator or prey.

It is only we humans who, under God's evolutionary guidance, have risen out of our animal roots to the level where we have the brain power and intellectual capacity to look at both what we have done, as well as what we are about to do, and understand that it is either good or evil. Out of all of Earth's creatures it is only we who have been led by God to this knowledge.

We have to balance that knowledge against all the instincts we have brought with us as part of our own evolutionary rise out of the unthinking animal world. For we too have to use the equipment evolution has given us to kill what we must eat, be it either animal or vegetable. We have no alternative other than starving to death like any food-deprived biological creature. Our evolutionarily provided tools, however, do not include claws or fangs; instead, we are bipedal creatures possessing only moderate speed and strength as well as dull teeth, but we are creatures who can grasp with our hands and think at exceptional levels in comparison with Earth's other species.

When pitted against the rest of world's life forms, it is our very anatomical limitations that have forced us to use alternative means to offset our physical disadvantages. We do so by engaging the intellectual abilities inherent in our brain capacities. This is one of the necessities God has created to drive us towards becoming what we now are. That includes our ability to recognize good and evil, even as the animal instincts we've brought with us out of our evolutionary heritage shout for either individual fight or flight in all our threat confrontations.

Historically *Homo sapiens,* like the rest of the animal kingdom, have killed to exist, to obtain food, and to clothe our bodies. We have fought our own kind to gain or control territory we felt we needed for sustenance, safety, or power. We have sought to dominate those around us as well as our environment. We have even found pleasure in killing for the sake of killing. This pleasure is not unlike that of a fox in a hen house that kills far more than it can consume, or a wolf or coyote slaughtering more sheep from a flock than it can drag off and eat. In short our ancestors' conduct mirrored much of the rest of the animal world even more than we do so today.

In terms of differentiating between right and wrong, our pre-hominid line of ancestors was no more capable of cognitively recognizing either the beneficial or the harmful impacts of their conduct than any of the other life forms surrounding them. That began to change, however, as our *Homo* line evolved into humans and became more and more intelligent. With these evolutionary advances something else was added. Under God's guidance, they began to intuitively feel and see that there were aspects of their actions that had implications beyond just survival and procreation. As their conduct went beyond what was required to meet their basic needs for food and shelter, they began to understand, if only dimly, that certain of their actions were wrong, while others were good in terms of their impacts beyond the individual actor.

And with this recognition good and evil sprang into existence where they had not existed before. Prior to that point only the natural order of the animal world and the physical universe itself had existed, at least here on Earth. Before then, good and evil had no relevance or meaning to the natural order of things.

With their ever increasing intellectual capacities, however, our human ancestors began to think and act beyond their immediate mates
and offspring. As part of their animal heritage they began to compete
with each other for dominance of their immediate multi-member
groups, as well as any other human groups they encountered. In doing
so, they had to choose between sharing the food and other resources
available to them or hording those resources for themselves and driving off others who needed them as well. They had to decide whether
or not to abandon their sick or elderly group members and those who
had been injured in the hunt, confrontations with other bands or by
some natural calamity. As they came to recognize these dilemmas,
they were forced to think and make what amounted to ethical choices
based on the challenges God had confronted them with.

The animal instincts they brought with them on this evolutionary
journey pointed them in the direction of taking what they wanted without any thought of sharing even if it meant killing or letting others die
to get it. Those instincts drove them towards getting and keeping as
much as they wanted of whatever they desired no matter who or what
they had to hurt in the process. This heritage was towards individual
self-protection and self-gratification at any cost.

Faced with this evolutionarily inherited challenge, God has had
to spend millions of years dragging our evolving line away from its
inherited animal instincts. Those instincts being selfishly focused on,
first and foremost, individual self-gratification. To overcome these
proclivities God has had to lead us towards a more nearly perfect understanding of what our conduct means. This required God to teach us
to recognize when our actions crossed the line from legitimately meeting our needs to bordering on excessive behavior and finally to ultimately being greedy and evil.

In the Biblical sense, a loving God has given us to eat of the tree
of knowledge precisely so that we could discern the differences between good and evil, right and wrong. God needed to do so because,
contrary to the mythological story of Adam and Eve found in Genesis,

80

humanity has never existed in an original state of perfection from which it fell into evil as the original sin.

Instead, as part of the rise of our hominid line, humans have passed from a purely imperfect animalistic state in which there was no evil – because our lineal evolutionary ancestors were incapable of recognizing their conduct as such – into a new state where our eyes have been opened to the possibility of evil. It was at that first moment when the first hominid experienced a glimmering of such an understanding that the concept of evil, and evil itself, sprang into being. It is in this sense that the biblical story of Adam and Eve's eating of the tree of knowledge is metaphorically viable.

From that point humanity's ever expanding growth in both population and occupancy of the Earth has been accompanied by an ever expanding scope for the possibilities for evil actions. With the evolutionary growth of *Homo sapiens*, not only did our intellectual and analytical abilities burgeon, but so did our capacities for both individual free will acts and concerted conduct in massed groups under the direction of hierarchical leadership. However, the dark side that accompanied these advances was our exponentially expanding proclivity for acts we were capable of recognizing as being immoral conduct.

With an ever growing comprehension of what they were doing, our ancestors began to use their increasing intellects to knowingly create more and more forms of evil conduct. They did so in order to satisfy their own personal whims, their greedy desires to possess material things at the expense of others, to inflict pain and instill fear in the peoples around them, or to enforce submission to their will: in short, to gain power and possessions at the expense of others.

This was more than just killing so they could eat as the animals do; instead, it required much more complex conduct based on conscious thought and effort as well as on the dark side of human imagination. It was done in order to produce the desired effects on those their actions were directed at. Under the emotional lashes of greed, apprehension, anger, hate, lust, envy, retribution, or desire for power, as well as other similar forces, our ancestors worked out a host of ways to induce fear and pain in those they decided they wanted to dominate, punish, or just torture. These actions were especially evil when they were done out of a naked desire to achieve selfish personal satisfaction from the infliction of all this harm on others. Thus did humanity knowingly create and recognize evil.

Using God's silent voice the Deity, in anticipation of this problem, has spent hundreds of thousands, if not millions, of years patiently teaching us the antidotes to these proclivities: love and morality! Starting with love for their mates and offspring, God has quietly pushed outward the realm and reach of the love humans and their evolutionary forbearers were capable of encompassing and sharing to an ever increasing circle of our fellow hominids, as well as to creation itself. At the same time God has been teaching us to love God's own self and the meaning of moral conduct. But the doing of all this first required our recognition and understanding of good and evil. Why? Because without such comprehension, love and morality would have had far less meaning for us; in its absence we would have had much less experience and understanding of the challenges our own natures presented to those two forces.

And even though God's focus has been on teaching us both love and morality, we need to constantly keep in mind that, though related, the two are not just versions of the same thing. Love flows from our hearts and sentiments reaching out to those we love. It is more an emotion and a feeling than a process of thought. Morality is far less a feeling. Rather, it is far more of an intellectually produced attitude, even if it is in many ways based on an underpinning of love. Morality provides the basis upon which humanity has been able to build the best of its societies. It creates the lattice work that facilitates society's capacity for living together in something approaching harmony and caring.

God has been instructing us in both love and morality because God not only wants us to outgrow the animal instincts that have drawn us towards conduct we can now recognize as evil, but also because God is love based on God's morality. God wants us to mirror and reflect God's own self as far as we humanly can. Additionally, it is reasonable to assume that God has gone to all this patient effort to create both our universe and us because God has plans for us that require us to be far more nearly perfect than we now are. And without absorbing God's lessons of love and morality we will never overcome evil and reach the levels of near perfection that will allow us and our souls to weather the ultimate end of our physical universe, while continuing to serve God in whatever capacities God has in mind for us.

So rather than having forbidden us – as proto-typical Adams and Eves – to eat of the fruit that leads to an understanding of good and evil God has bent every effort to ensure that we absorb that knowledge so that we can become a little more like God's own self.

Chapter Twelve

STORIES

It appears that God has chosen stories as one of the major tools for fostering humanity's growth out of our evolutionary kill or be killed animal heritage and seems to have done so because of a recognition of our ability and proclivity to learn from such stories. Reaching back into dim antiquity and in every human culture on Earth, we find a tradition of storytelling as a principal teaching vehicle for passing down knowledge from generation to generation.

How far back is that? One only needs to watch the 2011 Werner Herzog film *Cave of Forgotten Dreams* for a pretty good suggestion as to the answer. This movie takes us deep within the Chauvet caves of southern France. Coming out of absolutely pitch black darkness, the viewer is suddenly confronted with incredibly vivid drawings of lions, bison, bears, horses, and wooly rhinos. They seem to leap off the cavern's walls. All of these animal depictions were done by ancient humans over the course of about five thousand years, between roughly 32,000 BCE and 27,000 BCE.

Other than these drawings there are no other traces of human habitation within Chauvet's precincts. It is as if these underground lairs, along with the cave art in other locations such as Tito Bustillo in northern Spain and France's Lascaux were, in effect, humanity's first attempts at temples, albeit subterranean ones in nature. Looking at their lifelike representations of fearsome beasts in what feels like motion, you know that there had to be stories and rituals associated with them.

The pre-historic artists who made their way into the bowels of the earth or the shamans who accompanied them must have told those stories with spellbinding intensity and detail. And after stumbling through the blackness of those caves, confronting these true to life renderings of such powerful and dangerous creatures must have seared those tales into the memories of their awestruck and disoriented listeners. Those cave dwellers and all the subsequent participants in such descents into mystery would have passed on those stories and the information they conveyed to all the rest of their clans as well as the generations that followed them.

As part of these tales our ancestors would have certainly tried to grapple with their dimly emerging notions of the other worldly powers they felt surrounding them in all the things they didn't understand. Their attempts would have led them to ask questions of both God, in whatever form they conceived the Deity to be, and of one another – just as we do.

God would have responded with God's silent voice. That voice was just as hard or harder for those ancients to hear as it is for us to do so today. Their human interlocutors, however, answered out loud. Those answers would have often taken the form of stories including the stories that must have grown out of or inspired the Chauvet drawings.

Humans use stories to make sense out of our world. They do so because they have an innate need to understand not only the what of the world but also the how and why of it. This is one of the things that separates us from the rest of the animal kingdom.

Recognizing this, God patiently infused into humanity's endless supply of tales the thoughts, ideas, and commands God wished us to absorb. Even though influenced by God's silent voice, these stories were and are based on what the storytellers saw and experienced in their lives, including the stories they inherited. They also were limited by each teller's understanding of the world surrounding them, its inhabitants, and the universe beyond.

Within this framework, they and we have filled such tales with our perceptions of God and God's answers to humanity's questions, as best the questioners heard them. Additionally they contain our own

85

human speculations and hypotheses as to what those answers should be and what they mean; however, down through the ages people haven't been able to resist the temptation of declaring that their human infusions into God's answers came purely from God and not from the humans claiming to speak on the deity's behalf.

Historically these stores have come to be one of the principal ways we have communicated precepts, values, and instruction to each human generation. It is through them that we have learned so much of what we understand God wants of us. In fact, even today much more of what sticks with us in terms of values and expected conduct comes from such stories, as opposed to learning out of books of law or dry recitations of ideas and rules.

When you think about it, there are probably very few people who have ever actually opened a law book, or for that matter a book of philosophy. The only "law books" most folks ever physically touch, let alone read, are the Torah, the Bible, the Koran, the Upanishads, or the teachings of Buddha. And when people read or listen to those books being read aloud what they remember best from them are not unadorned rules, but the stories that embody those rules, giving meaning and context to them.

This is especially true now. We watch the nightly news broadcasts, read the daily papers and weekly magazines, or surf the net for various blogs to focus on the stories they contain. We pour over history books and fiction, watch movies and TV shows, all the while picking up on conduct, both good and bad, from the tales being presented to us through such media. One way or another we learn from all of them whether they are based on fiction or fact.

The common denominator in all this, what sticks in our minds, are the tales that carry a message. We likewise learn from the often times painful lessons of our own personal or financial stories and those of the people closest to us. In fact, we absorb most vividly and deeply the lessons we learn from actual experience as our own life stories unfold before us.

Obviously, stories are not the only way we learn, but they are certainly among the principal ways we acquire non-scientific knowledge and standards of behavior. God, having elected to work through the natural world, has focused on this trait. In doing so, God's silent voice has induced generations of storytellers to weave God's instructions and message into the fabric of their tales. God then fosters the perpetuation of those tales God approves of, and suppresses those God disfavors. Utilizing the silent voice we've been discussing, God tells both storyteller and listener that this story is to be remembered and repeated or that one can be forgotten.

However, as part of all this, God has had to accept the intellectual and knowledge-based limitations of both the story tellers and their listeners. For example, it would have done God no good to have silently attempted to teach the ancients about cosmology, Darwinian evolution, and the physics that produced their world over billions of years based on ideas and numbers they couldn't have comprehended. The science and math would have been gibberish to them.

But God had to give them some sort of an answer when they asked about, say, the origins of mankind and our world. So God allowed the fostering of the Adam and Eve in the Garden of Eden story and the other tales we find in the Book of Genesis and similar writings. Within those stories God embedded the concepts of knowledge of good and evil, of punishment that flows from violation of God's commands, and of choosing whether we do right or wrong.

We now know that there never was a time just a few thousand years ago when there were only two human beings on Earth living in idyllic tranquility. The idea of a very first otherwise perfect man and woman, whose fall from grace created that original sin out of which poured all subsequent human transgressions, is pure myth. Based on evolutionary science we can now be confident that, contrary to this tale, our human heritage is the result of billions of years of evolutionary growth out of an imperfect animal world to our less imperfect humanity of today.

Nonetheless, we are still captivated by the idea of an Adam and Eve living in that perfect Eden depicted in the first chapter of Genesis. It has flowed down to us over the millennia and, until modern times, was almost always taken to be the gospel truth – a truth derived from a story that has formed the basis of much human thinking.

Another prime example of this God-induced phenomena is the story of Noah and his Ark found a little later in Genesis (Chapters 6 through 9.) It is a story of such power that even our children are fascinated by it. Down through history people have been endlessly drawn to the tale. Even today groups have mounted expeditions to find the remains of the Ark and its ultimate resting place. Replicas of the Ark have been built in locations as diverse as the Netherlands and Williamstown, Kentucky.

Since many people take this biblical story to be literal inerrant history, let's examine it in detail. In doing so we need to keep in mind, however, the milieu and setting those who wrote it down all those thousands of years ago existed in. They inhabited a small area of what is now the Middle East. Their principal modes of locomotion were walking, riding animals, or sailing on small and not terribly sea-worthy vessels. These ancient authors would probably never have even suspected the possibilities of earth's huge size and planet girdling oceans separating massive continents and towering mountain ranges such as the Himalayas and the Andes with all their indigenous flora and fauna. They would have come in contact with only a small fraction of the life forms spread all across the vast expanses of those lands and seas, and they couldn't have comprehended the tectonic forces that over hundreds of millions, if not billions, of year shaped all that immensity.

So in this setting they crafted a mythic tale of God, angered by the depravity of human kind, electing to wipe out all of humanity and starting over. The only exceptions to this genocidal death sentence were Noah and his family. They were the only people on Earth God judged to be righteous enough to be worth saving.

In this tale, God informed Noah that God would shortly bring on a worldwide flood which would submerge all the Earth's land masses destroying all terrestrial life. To survive this deluge Noah and his family were directed to build an ark 300 cubits long, 50 cubits wide and 30 cubits deep (477 feet by 72 feet by 44 feet). Into this Ark Noah was ordered to cram breeding pairs of all the Earth's land-dwelling creatures plus the fodder and food necessary to support both them and Noah's family until the flood subsided.

Faithful to God's commands, Noah and his family of seven set to work building this vessel with three separate decks. When the appointed day drew near, God gave Noah seven days warning to load on board examples of all of Earth's creatures containing even the tiniest breath of life. Then God caused "all the fountains of the great abyss [to] burst forth and the floodgates of the sky were opened."[21] In combination this inundated the entire world until all its land features including its tallest mountain (what we now know as Mount Everest) were submerged by this catastrophic flood. And all terrestrial life, not on the Ark, was wiped out.

God's rain fell for forty days and forty nights. Once that stopped, the waters maintained their height (or depth if you prefer) for one hundred fifty days thereafter, receding over another one hundred fifty days until the Ark grounded on the re-emerged land being exposed by the falling waters.

Once the Ark had lodged itself on that land Noah waited a further forty days before releasing a raven and a dove to see if the land was dry enough to disembark the Ark's menagerie. But finding nowhere to alight, the dove alone returned. So Noah waited seven more days and then sent the dove out again. This time the bird finally returned with an olive leaf in its beak indicating that there was dry land that would support the Ark's population.

Based on this sign, after a further seven days Noah discharged his Ark's load of creatures to repopulate the entire world once again. And finally, as heavenly proof of a covenant between God, Noah, and his descendants that God would never again utilize a flood to exterminate all life, God created and placed in the heavens the rainbow.

Noah's flood story, however, is not unique among the vast repertory of story lines found among the various human cultures that have populated the world. Versions of the tale of a great deluge have been extant since the earliest known Middle Eastern writings. It is found inscribed on the clay tablets of Mesopotamia where the great god Enlil is described as having once ordered up a cataclysmic flood because humanity had become so noisy that they were disturbing his sleep, so he decided to do away with them.[22]

It is also found in the epic Babylonian saga of Gilgamesh and his search for immortality. During this quest he meets Utnapishin who'd been warned by the water god Enki of an impending flood that would devastate the Earth. This disaster was being visited upon all mortal life by the rest of the gods. In response to this warning Utnapishin built a boat in which he saved not only his own family, but also the artisans and animals needed to sustain both life and human culture; likewise, the Greek and Roman civilizations had their tales of Deuclion and Pyhrea who herded animals and their families onto a great box-like vessel to survive an inundation.[23]

Nonetheless, out of all these ancient flood stories only one has remained widely extant in the minds and literature of human beings: that of Noah and his Ark. The other ones, such as the tale of the peeved god Enlil who wanted to wipe out the raucous clamoring's of humanity so he could get some rest, have faded into nothing more than archeological footnotes.

Under God's guidance, however, Noah's tale moved robustly onward until it finally became part of the Book of Genesis. Within it God's silent tutelage embedded concepts God wished to impart to us, including listening to God's silent commands even when we are scorned by those around us for doing so. Other messages found in this tale encompass a number of additional concepts: that there is divine punishment for doing wrong, salvation of the righteous, and divine love and forgiveness, plus most importantly that God is trustworthy but cannot be judged by the same standards that God uses to judges us.

However, as an accurate description of real historic events, Noah's story just doesn't "hold water." What it really is, is an incredibly memorable, but mythic, tale that has captured the imaginations of endless generations of humanity. As such it has been very effective as a teaching tool for God. But in reading it today we are not absolved from the responsibility of doing so in the light of modern science and knowledge. We cannot hide behind the limitations Genesis' original authors labored under, claiming that since this is what God told them all those thousands of years ago we have no choice other than to take it as absolute truth today.

A quick analysis of the story's declared facts will demonstrate both its mythic nature as well as, and most importantly, the memorability of its God given messages. Starting with the forty-day inundation that submerged even Everest; we can calculate that this would have had to produce enough water to raise worldwide sea levels to a height greater than 29,000 feet, the height of that mountain. If even half that water was rain produced, it would have had to come pouring out of the sky in the span of 960 hours (24 hours X 40 days).

This half would have required rain fall rates of three inches *per minute* or 15.105 feet *per hour* (15.105 ft. X 960 hours = 14,500 ft.). For comparison, just think about how heavy we consider a downpour of three inches *per hour* let alone three inches *per minute*. A deluge of three inches *per minute* or 180 inches per hour is not rain; instead, it is more like a worldwide solid wall of water pouring out of the sky everywhere all at once. It would have been as if the waters of Niagara Falls were pounding down on every inch of the Earth's surface for those 960 hours.

The force of such a fire-hose blast of descending water would, in all probability, have torn any wooden vessel's roof or covering off. That would have led to the flooding of the Ark and its sinking, if it didn't first tear the entire vessel apart. Just picture in your mind's eye all the videos you've seen of the floodwaters of raging rivers destroying wooden bridges and homes as they swept by. It would likewise have been impossible for the air-breathing creatures, not to mention Noah and his family, to find any air to fill their lungs under such a solid inundation.

Next we have the question of how Noah's family could have herded together pairs of all the Earth's animal life forms drawn from seven continents and innumerable islands and moved them to the Ark's location. They would have had to have been brought across vast oceans, over mountain ranges, and across great deserts. In the case of whales, dolphins, walruses, seals, and sea turtles, they would have had to have been hauled out of the sea since, just like the land animals, under the great deluge that was to come, those creatures couldn't have drawn their own breaths of air either.

Today biologist tell us there are approximately 5,490 mammalian species, 9,084 reptilian, 9998 species of birds and 6,433 amphibians (frogs, toads, salamanders, etc.) spread across seven continents and a vast array of islands. We also know of over one million distinct species of insects and 102,248 arachnoids (spiders). All that life in male and female pairs would have had to be crammed into one vessel, with only three decks, 477 feet long, 72 feet wide and 44 feet deep. For comparison this is far smaller than a modern day cruise liner or aircraft carrier.

Additionally Noah would have had to first find and then load on board the Ark enough food and fodder to feed all these creatures plus Noah and his family for at least 401 days. That is Genesis' tally of the total duration of the flood and the time Noah is reported to have waited before freeing all that wildlife. No vessel of these dimensions could have physically contained such a biological mass of creatures or that much food. It would have sunk as a result of being overloaded even if Noah could have gotten even a portion of this menagerie and the necessary supplies onboard.

Then we have the even worse problem of how all those creatures would have survived, let alone gotten back to their original land masses and habitats once they were released. The food chains they'd have needed to sustain them after leaving the Ark would have been entirely obliterated.

Having been totally covered by thousands of feet of water for, at the very minimum, at least a year or so the land that emerged from the receding flood waters would have been nothing but slowly drying mud, rock, sand, and a slop of decaying vegetable matter mashed to a pulp. All the vegetation that had previously provided these creatures their habitat would have been drowned, dead, and destroyed. Any bushes or trees still standing would have been bare, rotting wood.

There would have been no grasses for the herbivores to graze on, no leaves or other forage to eat. The insects would have found no flowers or succulents to seek out for nectar and food. There certainly wouldn't have been any prey to sustain the predators other than the pairs from the Ark that would have been quickly consumed before they could repopulate the world's wild herds.

Finally, there would have been no living olive tree from which Noah's dove could have plucked a living green leaf to carry back to Noah. All the Ark's animals would have died of starvation or been killed by the released predator species long before they could have confronted the problems of crossing continents and oceans to repopulate their original habitats spread all across the Earth's various land masses.

We also need to consider the rainbow that Genesis says was first created by God after the flood in order to serve as a perpetual symbol of God's promise and covenant to never again destroy all life through the real world mechanism of a flood. Simple physics teaches us that rainbows would have appeared in Earth's atmosphere long before humans evolved into existence. As soon as there was breathable air it would have contained water vapor as one of its constituent gases.

Today we know that when sunlight passes through the water molecules of that vapor, it is refracted and broken up into its spectrum of primary colors, thus producing the phenomena of the rainbow. This would have been occurring long before the advent of *Homo sapiens*. Noah would have certainly seen such bows from his infancy, so it couldn't have been a phenomenon first noticed after the flood.

Those who want to declare that every word of the Bible is inerrant and absolutely irrefutable fact may say that, well, God used God's logic-defying and limitless powers to first roundup all those creatures from the four corners of the world and then bring them to the Ark for Noah. Then God crammed all this life and the stores necessary to sustain it into extra-dimensional space within the Ark following which God protected the Ark itself from destruction by the falling walls of pounding water. They will argue that using those same powers God re-vegetated the entire Earth in an instant and provided meat for the released predators without their touching their fellow prey species passengers. Thereafter God transported them all back to their points or origin all across the face of the globe.

But as we've seen, more often than not God works through the natural laws and mechanisms of the world and universe to accomplish God's will. In fact Noah's story starts out with God using the real world phenomena of a flood produced by massive rainfall as God's agent of destruction. If God had to resort to God's other-worldly powers to complete God's salvation of all forms of life, it would seem far more logical and easy for God to have used those powers up front to do the dirty work of destroying life. All God would have had to do was reach out and disrupt the electrically based nervous systems of all the creatures God was dooming, thus stopping their hearts and respiration producing instant death. But that didn't happen.

What did happen is that God took a compelling fictional story and embedded into it a series of lessons that down through the centuries would be understood by all those who encountered the tale in any form. As we've noted, these lessons included several concepts: there is punishment for wrongdoing and salvation for the righteous; God loves humanity, especially those who follow God's commands, and life itself; humans need to listen to and follow God's directions because God's word and covenants can and should be trusted; and most importantly, we humans cannot judge God by the same standards God utilizes to judge humanity.

As to this last lesson, the act people don't seem to find shocking today is that our loving God is portrayed as committing genocide: wiping out all of humanity with the exception of only one lone family. And then at the end of the story that same God issues the command:

> ....I will demand an accounting for human life.
> If anyone sheds the blood of man,
> By man shall his blood be shed;
> For in the image of God
> has man been made.
>
> (Genesis 9:5-6)

This from the deity who had just been described as so grieved by humanity that there was godly justification for wiping it off the face of the Earth along with all other life and starting over again. Clearly then Noah's story teaches us that we cannot judge God even though God judges us.

When we subject any number of the Bible's other stories to this sort of scrutiny we can often find the same pattern of godly use in many of them. Thus, while they may not stand up to hard factual analysis, they are extremely memorable. And each contains one or more lessons God wishes to teach humanity in order to guide our growth and evolving conduct down through the ages.

Additionally, in examining each we have to keep in mind that at any given point in time God has had to work with the humans who were then available to God to act as storytellers or scribes, no matter how flawed they and their knowledge might have been. God also had to craft the stories God was fostering in light of what was the then human understanding of the world, science, chemistry, physics, mathematics, biology, philosophy, and all the other realms of knowledge we have come to take for granted today. All of these would have been far less comprehensible to God's audience when the stories found in Genesis were first written in any form, let alone verbally circulated down through the preceding generations.

It would not have helped God to suggest to the ancient storytellers trying to deal with the supposed history of Adam and Eve and of Noah, the actual time frames of human evolution that, as we now know, have been in the billions of years. Numbers of that magnitude would have been incomprehensible to those ancients who thought in terms of human generations of less than one hundred years. Those people obviously found it utterly amazing and nearly superhuman when they attributed life spans in the multiple hundreds of years to the heroes of their tales.

In crafting stories and lessons that would be understandable to those who first heard God's silent voice, God had to accept misconceptions and factual impossibilities within the immerging storylines so long as the core messages God wished to foster were being conveyed. Thereafter, when the time was right and human comprehension had achieved levels better able to grasp and absorb new ideas, God could begin the task of correcting those misconceptions.

The shift from the Old Testament portrayal of a wrathful deity dooming entire populations to that of the loving God found in the New Testament who offers salvation to all is a clear example of such a correction of focus and understanding under God's silent tutelage. We should not be surprised or upset by the possibility of God's use of imperfect or fictitious stories in this way. When you think about it, many – if not most of us –do the same thing with our own children.

To see just one example of our doing so we need go no further than Santa Claus. As a society and as individuals we happily perpetuate the fairytale fantasy of a magical and jolly old elf and his little elf helpers spending each year in the regions of the North Pole making toys that Santa then delivers to good little girls and boys all over the world each Christmas. He accomplishes this feat via a sleigh pulled by eight flying reindeer that can magically reach the homes of each and every child in just one night.

This, of course, is a fairytale, but it's one we all love and have done so since our own childhoods, just as our parents and their parents before them did. Parents and society find nothing wrong in teaching kids something they know is untrue; instead, we use it to inculcate concepts of wonder, joy, and fantasy in our children. It also becomes part of teaching them basic notions of reward and punishment for either right or wrong conduct and the consequences of being either good or bad little boys or girls. It starts them on the road of learning the happiness that giving brings. It conditions children's thinking to accept the possibility of there being events and realms of being that reach beyond the concrete realities of our everyday world and lives.

God uses stories and myths in just this same way to lovingly guide us on the paths of our growth out of selfish, greedy, and evil conduct. Religious institutions, such as the Catholic Church, have endorsed this conceptual fact. In a modern version of the Bible, specifically approved and authorized by the recently beatified Pope Paul VI, the commentary note introducing the Book of Genesis cautions the following:

"To make the truths contained in these chapters intelligible to the Israelite people destined to preserve them, they needed to be expressed through elements prevailing among that people at that time. For this reason, the truths themselves must therefore be clearly distinguished from their literary garb."[24]

Thus, we have now reached the point in human development where God, as the inspirational author of the Bible, Koran, and all the rest of the cannon of the world's defining religious writings, seems to want us to get past any need for literal truth in the stories contained in those works. God is telling us, through God's silent voice, that we should both absorb and live by the embedded lessons and teachings we take from these stories, without the need for such stories themselves to be factually true in every detail.

As the well-respected twentieth century theologian John Hick has written, "the truth of a myth is a practical truthfulness: a true myth is one which rightly relates us to a reality about which we cannot [adequately] speak in non-mythical terms."[25]

The issue for us now becomes what people do with these stories, including those that are myths, and their teachings. Unquestionably they form the foundations for much of our western societal and religious thinking. But, given these foundations, each individual must decide what will be built upon those stories and myths, generation after generation.

For many, all they can build from this base is a wall of impenetrable stone that blocks any further advance, any further thinking and analysis beyond the stories themselves. This wall becomes something they can't see through or past; instead, they are frozen, rooted rigidly in place staring at what, for them, have become inflexible imperatives inscribed on the surfaces of those stone-like stories and rules. They are all piled into a barrier in front of them. For these individuals this barrier constitutes God's written voice that they must take literally without any form of analysis or flexibility in interpretation.

For others it is possible to use these stories and their messages as the building blocks needed to construct a gateway that arises out of that underlying foundation. It provides a portal that leads them toward humanity's future growth. But each generation must reconstruct and pass through that portal on their own.

Thus, this is not a onetime process. The wall is ever with us and must always be considered. It is never behind us. The stories framing this gateway become a part of the guiding lessons God has so lovingly engraved on the mosaic formed by its stones. If we will look for their bigger picture, the portal will open for each of us. It will help to guide us toward what God wishes humanity to become. How we use these building-block stories is our choice. However, it is up to each of us to make free will elections as to what we build with them, solid wall or gateway.

But as we do so, we should always keep in mind where our cultures and societies and we as individuals would be had God not lovingly molded us over nearly endless generations. To do this, God used the positive changes generated by God's focusing us on the lessons embedded in the stories found in every philosophy or religion's teachings in ways that can be best understood by the adherents of that religion or philosophy. Without those godly efforts would we care as much about the poor and hungry? Would we have built as many institutions and enacted as many laws to protect the weak from the mighty? Would even more of us be prone to murder, lie, cheat, and steal? And would sexual license run more rampant than it currently does?

To achieve these positive outcomes God uses stories to change our thinking and perceptions of life in positive ways. This implies that God approves of and seeks such positive change in the human condition. Through these stories God has taught us how to recognize the abstract concepts salted throughout such tales of human conduct. They include the differences between good and evil, the concepts of love and caring, and the possibility of salvation and forgiveness. However, it is up to each of us, to each generation, to parse God's stories in order to work out the lessons they contain for each succeeding cohort. Once we do, each individual has to then apply them for the positive advancement of the individual and humanity.

Likewise, in seeing all that has gone before us, people need to recognize that God is forward focused. God has been patiently advancing humanity out of our past animal heritage, toward a more perfect future for humanity as a whole, and what we and our progeny can become individually under God's direction. God is not trying to return us to humanity's far from perfect past but, instead, is honing us for the future. God's focus in all this is the growth of each of us, individual by individual and society by society, toward a perfection we can achieve if we accept God's directions.

Chapter Thirteen

A LOVING GOD'S USE OF VIOLENCE

As we consider how our loving God has so patiently created our universe and shepherded humanity out of its animalistic past, each of us is confronted with the necessity of coming to grips with the fact that one of the tools God has used in this effort is violence. This includes violence in both the physical universe that encompasses our world, as well as violence in human conduct. However, even as we recognize God's use of violence, we need to refrain from jumping to the conclusion that such use means that God is less than loving or is to be censored by humans for employing violent tactics.

In Chapter Three we looked at how God created the universe through a long series of violent events starting with the Big Bang itself. That unimaginably violent explosion was followed by a universe-wide collision of particulate matter and antimatter in an antimatter annihilation event that left what became dense clouds of matter in the form of hydrogen. Those clouds in turn came together to form endless numbers of suns, as each dense and essentially solid ball of compressed hydrogen achieved the critical mass needed to erupt into violent thermonuclear fireballs that put human-built hydrogen bombs to shame. Those fireballs contained all the sustained explosive energy needed to light the universe and nurture life for billions upon billions of years.

After approximately 9.3 billion of those years, our own solar system coalesced as a part of the explosive ignition of our Sun. In that process the planetary ball we think of as Earth was pounded together out of swarms of matter captured by this new sun's gravitational pull. What became that ball's molten iron core and rocky mantle were shaped by violent tectonic forces even as they were being bombarded by cosmic debris and violent collisions with other planetary bodies, thus producing both our planet and its moon.

It is no surprise to us that Earth, in and of itself, is a violent and dangerous place with volcanoes, earthquakes, tsunamis, hurricanes, tornadoes, blizzards, floods, and all the rest of nature's destructive forces, yet under God's authorship our world was and is not too violent for life to emerge and flourish. Humanity's rise has been part and parcel of that greater evolutionary march of violence and is not separate and apart from it. This evolutionary history of advance has, with God's concurrence, been based on the endless cycles of mayhem demanded by the food energy requirements of much of the animal portions of that life. These needs have been met through the animal imperatives of eat or be eaten, as each life form sought to ensure its own survival and propagation. All of this has, in turn, produced the endless lines of forced mutation that have led to us and the biological world as it now exists; thus, God has demonstrated that real world violence can and will be used to achieve the outcomes God is aiming toward.

This truth has been similarly manifested in God's dealing with humanity. It can be found in the history and teachings of almost every religion and culture down through the ages. We've already looked at the ancient story of Noah and his Ark that was handed down through time to the Israelites who then incorporated it into the Book of Genesis; a story whose genocidal depiction of God did not seem to shock them or us.

Likewise, the rest of the twenty-four books of the Jewish canon, the Hebrew Bible, have innumerable tales of godly use of violence. We need look no further than the fates of Sodom and Gomorrah depicted in Genesis Chapter 19, or that of the Israelite King Saul found in the Book of Samuel.[26] In that latter story when Saul did not doom a conquered pagan city's king as God had instructed, God is depicted as violently stripping Saul of both his kingship and his life in order to pass the rule of Israel and its people to King David, God's chosen successor to that throne.

Even peaceful Buddhism owes much of its spread across the Indian subcontinent and Asia to a series of violent conquests. One of the great empires of history was that of Mauryan centered in what is now India. Perhaps its most successful ruler was Ashoka the Great (ruled 272 – 232 BCE). He was a brilliant military commander who crushed various revolts and, in a brutal war (262 – 261 BCE) that cost an estimated hundred thousand lives or more, conquered Kalenga. However, his experiences in that devastating and horribly destructive conflict engendered a deeply felt remorse in Ashoka that led him to embrace Buddhism and its nonviolent tenets. Thereafter, he devoted the rest of his life and his empire to the successful spread of the teachings of Buddha across much of Asia from Sri Lanka in the south northward to Afghanistan and northern Asia including Siberia.

Arguably, another more recent example of this phenomenon is, after the death of the Prophet Mohammed, the explosive seventh and eighth century CE expansion of Islam through military conquest. In little more than two hundred years that conquest took this new faith from a small localized religion founded in the Arabian Peninsula to a major God focused militant one covering most of North African, the Middle East, and parts of Europe. This was a religion and associated culture that during the worst of the Dark Ages following the collapse of the western half of the Roman Empire preserved and expanded much of human intellectual development outside of Asia especially in fields like philosophy, medicine, mathematics, and astronomy.

But the paradox of a loving God's use of violence is perhaps best illustrated by the formation and rise of Christianity. Its beneficent outcomes, and the geographic expansions of its teachings, were the products of very violent beginnings. They were first orchestrated through the life and death of Jesus of Nazareth and the later rise of the Roman Emperor Constantine three hundred years thereafter.

JESUS

Two thousand years ago Jesus was an itinerant Jewish rabbi preaching, teaching, and curing the sick and infirmed in the general areas of the Galilee, Judaea, and Jerusalem. His message, the one God sent him to convey to the world, was one of love, forgiveness, healing, and salvation. At its core it could be summarized by Jesus' response to a Pharisaic lawyer's question:

"Teacher which commandment of the law is the greatest? [In reply] Jesus said to him, 'You shall love the Lord your God with your whole heart, with your whole soul, and with all your mind. This is the greatest and first commandment. The second is like it: You shall love your neighbor as yourself. On these two commandments the whole law is based and the prophets as well."[27]

But under God's guidance, and despite this message, Jesus went out of his way to court the animosity of Israel's ruling elites and their Roman overlords. He intentionally and openly violated any number of the commandments found in the Mosaic Laws as well as their offshoots, all of which governed the Israelites. These transgressions included curing people on the Sabbath (a day of enforced rest); eating without first washing his hands; consorting with tax collectors, prostitutes, and other sinners; refusing to condemn an adulteress; and openly proclaiming that he had the power and authority to forgive sins when only God was supposed to be able to do that.

But perhaps the offense those elites – the Pharisees, Sadducees, Sanhedrin, and the priestly classes – found most offensive was his flirtation with what they considered blasphemy. Jesus hinted or allowed others to suggest that he might be the long awaited messiah. He did so while knowing such blasphemy was punishable by death.

Adding to all this, Jesus endlessly heaped verbal coals onto the heads of Israel's leadership. He repeatedly denounced these elites as hypocrites whose outward observance of the law masked their avarice, self-dealing, and disdain for the rest of Israel. They were like vessels, polished clean on the outside but coated with filth on the inside. He branded them as blind guides. In short, Jesus courted their enmity as he goaded them into attacking him. And this is precisely as God would have it, even though it was not exactly "brotherly love."[28]

At the same time, while tantalizing the ordinary classes of the Israelite populace with the possibility that he might be their long-awaited messiah, Jesus ensured they would never know for certain what the answer to that question was or exactly what his message meant. To fuel this confusion he preached in parables that obscured his meaning.

When his disciples asked him why he did not speak plainly, Jesus quoted the Prophet Isaiah to the effect that the people of Israel would look, but not see, and listen, but not understand with their hearts. Instead he was teaching the meanings of his parables to his disciples and followers so that, when the time came, they might carry his message outward from the nation of Israel to the greater gentile world.[29]

This was not because Jesus did not love his fellow Jews, which he did. It was because, had they come to believe in him as their messiah, they would not have been capable of sharing such a faith with the gentile world surrounding them. This attitude was not their fault; instead, it was the product of their several thousand plus year tutelage under God's silent voice. In the face of never ending adversity, that voice had conditioned the Israelites to abhor the pagan gentile nations that both surrounded and constantly menaced them, and they were true to that conditioning.

In such a setting Jesus was a threat to both the Jewish establishment and to Roman rule. He was calling for change. And that possibility made the existing order fear a revolt of the ordinary people. If it occurred, they were certain it would lead to massive Roman reprisals. And they were right, since that was exactly what happened when just such a revolt did break out nearly forty years after Jesus' crucifixion.

The Jewish leadership, as represented by the Sanhedrin, was terrified by this possibility. They told each other, "If we let him go on like this, the whole world will believe in him. Then the Romans will come in and sweep away our sanctuary [Jerusalem's Temple] and our nation."[30]

As the final straw, just before Jesus celebrated his last Passover, he entered Jerusalem passing through the rejoicing crowds riding on an ass' young colt. In doing so he was fully aware of the Prophet Zechariah's proclamation, made some five hundred years earlier, "See your king shall come to you, a just savior is he, Meek, and riding on an ass, on a colt, the foal of an ass."[31]

With that the throngs started joyously hailing him as king. This of course drove the Jewish establishment into frenzy, just as God intended it should. The Pharisees demanded Jesus rebuke his disciples for allowing people to salute him in this way, but Jesus flatly refused to do so.[32]

Then even though he knew that the multitudes, and even his own disciples, wanted him to proclaim himself their king and messiah, Jesus intentionally left them hanging. An earthly kingship was not God's purpose for him.[33] This refusal caused the volatile masses to begin to swing against him, when the priests' and Sanhedrin's fear of the people had previously been his best protection against being arrested or stoned.[34]

Despite all this Jesus then placed himself in the now famous Garden of Gethsemane on the slopes of the Mount of Olives. It was not a garden at all, but instead a patch of wild land in an isolated position outside of Jerusalem's walls. There his foes could arrest him at night without fear of a clamoring crowd violently opposing their efforts. This led, as Jesus knew it would, to his trial with its foregone judgment of blasphemy since he offered no defense. Of course this meant his death in one of the most publicly barbaric forms of execution the Roman world could devise.

First Jesus was scourged by Roman legionaries so that he would be too weak to effectively resist when he was nailed to a cross. Then his cross, with him on it, was erected in a very public place where both Jerusalem's inhabitants and, more importantly, Jesus' disciples could watch him slowly die in excruciating pain. There he hung suspended, his extended wrists pinned by nails to the crossbar of this instrument of torture and his hips and ankles twisted sideways with a single spike driven through both shinbones into its post. In this contorted posture his body's weight pulled agonizingly against those nails. And here he fought against both awful pain and creeping asphyxiation, induced by his horribly twisted and unnatural body position, that all combined to slowly constrict his lungs.

All this led, as was intended, to a slow death of extreme suffering. But to confirm the fulfilment of Jesus' death sentence, when Jesus finally expired, the Roman soldiers supervising his execution thrust a lance into his side just to make sure they had put an end to his life. Watching this, all the onlookers were finally sure he was gone and the crowd slowly dispersed leaving his body hanging there like a rag doll for some time thereafter. It was only taken down toward the end of the day when the Roman procurator of Judaea gave permission for the corpse to be removed and buried in a tomb hewn out of rock. Then, with the body finally inside that sepulcher, it was sealed by a boulder rolled across its entrance.

But if Jesus was the Messiah, the Christ, sent by God to teach a new way of worshiping God as well as how to interact with other humans in order to merit eternal salvation, why would God allow him to be cut off in his prime and suffer such a horribly humiliating and violent death? Why not permit Jesus to live on to a ripe old age, imparting his message to more and more of the Jews he lived among? What purpose did a loving God achieve through all this agony and violence?

The answer is that if Jesus had lived out a full life within Judaism we wouldn't have Christianity today; instead, Jesus probably would have been ranked along with Elijah and Elisha as one of a troika of Jewish prophets. Among them they would have been credited with raising people from the dead, curing the ill, and multiplying meager

107

supplies of bread, flour, oil, and fish to feed far more people than it would have seemed possible, given what they had started with. And the three would have been said to have been taken up into heaven without dying. In short, Jesus would have been credited with doing nothing more than one or the other of his two predecessors, Elijah and Elisha, were said to have done.[35] In this setting Christianity would, in all likelihood, have become nothing more than a minor sect within Judaism. It certainly wouldn't have become the world force it is today.

To avoid this outcome God both needed and engineered Jesus' execution in the prime of his life. It had to be a very public and brutally humiliating death. A death that was so certain that all who witnessed it, especially his disciples and other followers, could have no doubts whatsoever that he was dead and buried.

When that certainty was shattered by his resurrection, it produced such a profound impact on those who believed in Jesus that it transformed them. They went from the hesitant and fearful Jewish country rabble they had been, to confident and forceful leaders of what was to become Christianity. It also gave them the strength to withstand the trials they would, in their turns, endure. It gave a loving God what God required in order to begin the propagation of Jesus' message of love and salvation outward from the clannish and inwardly focused Israelite nation, to the greater gentile world the Jews had been so effectively taught by God to shun.

In a similar manner God utilized violence to drive Jesus' followers, and along with them the message and teachings God wished to disseminate, out of the insular world that two thousand years ago was Judaism. A loving God intended these teachings to be shared with the non-Jewish peoples surrounding Israel.

To do so, however, God had to overcome the xenophobia God had worked so hard to instill at the core of the Israelites' being. Having spent the preceding several thousand years drumming a rigidly inward focus into God's Jewish people – along with an absolute abhorrence of gentiles, and their false little gods of wood and stone – those people weren't predisposed to share beliefs with pagan gentiles. God had done this in order to preserve Judaism's central theme of there being only one true God, which was God's own self. But, by the time of Jesus, the rest of the Roman world had grown to the intellectual and spiritual point that significant portions of its populace were primed to receive a loving God's new teachings as proclaimed by Jesus.

As we've noted, after Jesus' crucifixion and resurrection, his initial adherents were almost all Jews. They still centered their thinking, worship, and activities within the confines of Jerusalem's Temple, Jerusalem itself, Judaea, and the Galilee. Absent God's goading they would have been content to continue that focus as they expectantly awaited Jesus' second coming as the Christ, accompanied by God's end times. They were in effect just another sect of Judaism and a very small one at that.

To overcome this God had to once again utilize the Jewish establishment to do God's work in the pushing. Having eliminated Jesus as a potential rebel threat, and seeing his followers as nothing more than another small sect of Judaism, the surprise is that the Jewish leaders were not content to just let things be. They were certainly familiar with the phenomena of Judaism's fragmentation into competing groups, parties, and sects. There were dozens of these; the best known being the Sadducees, Pharisees, Zealots, and Essenes.

All of them competed with each other for ascendance within Judaism's closed little world. In fact, in many ways they shared various conceptual beliefs with Jesus' followers. The Sadducees were the most rigid. They associated themselves with the priestly classes and denied the possibility of an afterlife, as well as the existence of spirits, either angelic or demonic.

The Pharisees by contrast, were far more adaptable and accepting of change than the picture of them portrayed in the New Testament. They practiced a simple way of life, as well as concern for their fellow Jews. They also integrated humane concepts into their legal interpretations of the Law of Moses and espoused a belief in an afterlife. In fact, a number of Pharisees seemed to have actually supported Jesus.

The Zealots, and the beginnings of what later became known as the Sicarii or knife men, openly opposed Roman domination of Israel. Both groups had strong messianic tendencies and the proto-Sicarii would, from time to time, assassinate Jews who collaborated with Rome or participated in the Hellenization of Judaea.

While likewise extremely messianic, the last of these groups, the Essenes were monastic or semi-monastic in organization. The majority of Essenes eschewed marriage and limited their membership to adult males. They proclaimed a strict form of predestination and had an emphatic end of days turn to their beliefs. Finally, they also rejected the reigning priestly class and the temple worship practiced by that class in Jerusalem as having corrupted the Jewish faith.

Clearly, divisive groups and sects within the Judaism of Jesus' time were nothing new. The Jewish establishment either tolerated, or at least didn't seem to seek to stamp them out, even though the Zealots, Sicarii, and Essenes were in open opposition, in one way or another, to the status quo of priestly dominated temple worship under Roman auspices.

Thus the question becomes why didn't the Jewish governing classes likewise tolerate this minor new sect that sprang up around Jesus' memory after the crucifixion in the same way they did the other sects and groupings? Why did they instead quickly and violently seek to suppress Jesus' New Way? One answer to those questions that springs to mind would be that using God's silent voice God urged the Jewish leadership to attack Jesus' followers. In heeding those silent urgings, the Sanhedrin thought that they were doing God's will, which they were, but without understanding its true purpose.

So attack they did. They first used stern warnings, arrests, and whippings of the Jesus sect's leadership. When these had no effect they escalated the charges to blasphemy. The punishment for that crime was the stoning to death of one or more of the group's leaders. This was followed by a general persecution of Jesus' adherents. Up to that point most of these followers of the New Way had remained in Jerusalem in order to continue worshiping as Jews in the Temple. There they proclaimed Jesus as the Messiah and limited their dissemination of God's new message to their fellow Jews, as they awaited Christ's eagerly expected second coming.

PAUL

The pressures produced by these relentless attacks eventually worked God's will. They forced Jesus' followers, now in fear for their lives, to flee the city. As a result they dispersed throughout Judaea, Samaria, and ultimately to the surrounding gentile nations. And of course, Jesus' loving message went with them as part of their flight, just as God intended it should. The major exceptions to this pattern were the Apostles themselves who, along with Jesus' brothers, stubbornly hung on in Jerusalem trying to remain part of the Jewish faith.

Obviously frustrated by the unintended consequences of their failed effort to strangle this new sect, the Sanhedrin began commissioning some of their more zealous advocates to pursue Jesus' followers into the gentile world. These men were ordered to root out the fugitives and charged with the responsibility for countering any successes the new teachings might have with the Jewish communities scattered throughout the surrounding nations.

One young man so commissioned was Saul of Tarsus whom we know as Paul. He was a perfect choice for the job. A strictly observant Pharisee and extensively trained in the Law of Moses and its practices, he was either a rabbinic lawyer or nearly so. He was also a Hellenized Jew in that he had been raised in both the gentile city of Tarsus as well as Jerusalem; thus, he was fluent in both written and spoken Greek, Aramaic, and Hebrew. He was also a citizen of Rome by birth. As such he could comfortably move about in the gentile world in which the adherents to Jesus' New Way had been forced to take refuge. So the

Sanhedrin sent Saul off to Damascus to uproot this heresy, never realizing that all this also made him perfectly suited to God's plan for spreading Christianity throughout the Roman Empire.

Bent on executing his mission of persecution, Saul had nearly reached Damascus when he had his fateful encounter with God's voice. In Saul's case that voice apparently rose close to or reached an audible level of intensity. God had selected him to take Jesus' message to the gentiles of the Roman world. It was a voice that violently drove Saul to his knees and left him temporarily blind as he writhed under its lashing commands.

Guided thereafter by God's silent but compelling instructions, the man who now called himself Paul spent the rest of his life trekking by land and sea back and forth across the eastern Mediterranean provinces of the Empire in a series of missionary journeys. He went from city to city where, with few exceptions, a pattern would unfold. Paul and whoever his companions were would first approach the local Jewish community through its synagogue. There he would proclaim Jesus and preach his message. And in each, under the guidance of God's silent voice, the majority of his Jewish audiences would reject what he was telling them. As they did, they felt they were doing God's will, which, in fact, they were.

In the face of each such rejection, Paul would then turn to the gentile population in that locale, and in almost all of their towns and cities he began to have success in terms of winning over some of the pagans who heard him to the teachings and position he was espousing. This, in turn, would infuriate that city's Jewish population. Even though they had previously ignored all sorts of other pagan or heretical religious ideas being fed to those same gentiles, the local Jewish communities would react violently to Paul's message. Time after time these enraged Jews would either attack Paul or incite their local city authorities to do so. Time after time they would attempt to silence him through physical intimidation or drive him out of town.

In 2 Corinthians 11:25-33, Paul attested to the mental and physically violent sufferings he had to endure during his missionary efforts when he wrote of:

"....my many more worse beatings and frequent brushes with death. Five times at the hands of the Jews, I received forty lashes less one; three times I was beaten with rods; I was stoned once, shipwrecked three times; I passed a day and a night on the sea. I traveled continually, endangered by floods, robbers, my own people, the Gentiles; imperiled in the city, in the desert, at sea, by false brothers; enduring labor, hardship, many sleepless nights; in hunger and thirst and frequent fasting's, in cold and nakedness. Leaving other sufferings unmentioned, there is that daily tension pressing on me, my anxiety for all the churches."

Using that silent subliminal voice, it would seem God orchestrated all these events. God did so not to punish Paul, but instead to keep him on the move. It was that pattern of movement which spread God's message of salvation and brotherly love to city after city all across the eastern half of the Roman Empire. In each, Paul would find a few Jews receptive to the idea of Jesus as Messiah plus a significant number of accepting gentiles. Together, that small core of interested gentiles and Jews were enough to form a viable local church, but the opposition was always strong enough to prevent Paul from settling down in any given local. It forced him to keep moving on.

Nonetheless, each little church he left behind weighed on his mind. He feared for their survival and the threat of those fledgling Christians being led into false doctrines and conduct by others. Out of this came Paul's letter writing campaigns that, either directly or indirectly, have given us thirteen books of the Christian New Testament and served as some of the bedrock upon which the Christian Church has been built.

By the same token God intended the majority of the Jews who heard Paul to reject him and his message. God did so not because God wanted to punish or disown those Jews. God did so because God knew that, given the choice, Paul would have been more comfortable among them. That's why Paul almost always started out in their synagogues when he came to a new town. Had they accepted Jesus as Messiah, Paul would have stayed in their midst. That was not what God needed; instead, God wanted to force Paul to turn to the gentiles to convey God's new teachings to a much wider audience. The Jews were only doing what God intended them to do in order to effectuate God's plan.

So this was as God planned, based on love, for it to be. It gave the world what it needed and was ready for: Christianity in a form that has been a major impetus to the growth of humanity out of our animalistic past, toward the future God intends for us. However, in doing all this, God did not hesitate to employ violence when it was the best way to achieve God's purposes and accomplished God's ends.

CONSTANTINE

We do not need to trace every possible instance of this godly use of violence through the ages, but certain key examples do bear special attention. The first of these is the history of one of Rome's most important emperors, Flavius Valerius Aurelius Constantinus, better known as Constantine the Great. His impact on the rise of Christianity is nearly incalculable, yet it is little noted or understood by most ordinary followers of the Christian faith much less the rest of humanity today.[36]

By late in the third century of the Common Era, Christians, though an often despised and occasionally persecuted minority had grown in numbers till they constituted what has been estimated to be something in excess of ten percent of the Roman Empire's total population. This had occurred during times of extended instability, civil war, and sometimes near anarchy. In just the forty-nine years between 235 and 284 C.E. the

Empire ran through twenty emperors with each change ripping at its fabric of existence.

At the end of this period a Roman general from the Balkans named Diocletian took the throne. When he did, the Empire was an utter mess. To regain control of this vast amalgamation of nations and peoples spreading from the Atlantic in the west to the borders of Persia in the east, including North Africa and Egypt, Diocletian cut the Roman world into eastern and western administrative halves. Even though Diocletian remained the Empire's paramount ruler, he placed one of his fellow generals Maximian in command of the west, while Diocletian focused on the east. Each general assumed the title of Augustus and each had a deputy ruler known as Caesar. The two Caesars, it was assumed, would succeed them as Augustus. The Caesar in the west was a third general, Constantius, and he had a charismatic son named Constantine.

As Diocletian battled to restore cohesion and control of this far-flung empire, he turned on the small but widespread Christian minority. He felt its belief in just one god, as opposed to the many gods of Rome, was what he considered to be an atheistic threat to the gods he believed in and what he was trying to accomplish. In his view the Christian refusal to acknowledge the official gods of Rome constituted a challenge to the Empire-wide public order and stability he was in the process of re-imposing.

So in 303 C.E. Diocletian decreed what was to be the last and one of the fiercest attempts at suppressing the Christian cult it ever experienced. This persecution was carried out with extreme rigor in the east. There, all avowed Christians faced torture and execution if they refused to renounce their beliefs. But in the western territories of Gaul, Germania, and Britannia, where Constantius ruled as the western Caesar, Diocletian's edict was never actively enforced with any real vigor.

However, Diocletian's exhaustive efforts to get his realm under control took its inevitable personal toll on the emperor. In 305 C.E. after twenty-one years of ruling the Empire, he was a worn-out sixty year old, and he decided to retire, abdicating the throne. But as part of doing so, he also forced his co-Augustus, an unwilling Maximian, into retirement with him. They were replaced by their two Caesars, (Constantius and an eastern general, Galerius), who now became co-Augustus. Unfortunately, this orderly succession fell apart when a year later Constantius died. With his death, the soldiers of his western legions took matters into their own hands. They proclaimed his son, the charismatic Constantine, then only in his twenties, as the successor Augustus in the west.

That didn't sit well with the still living and unhappily retired Maximian. At first he supported the claim of his own son Maxentius to be Augustus only to then fall out with him and flee to his son-in-law Constantine. When that didn't work out to his satisfaction he next led a failed rebellion trying to size the western command from Constantine and was captured near Marseilles in 310 C.E. resulting in his suicide.

All this left the young Constantine no alternative other than to prepare for what would be a bloody civil war. In the midst of that process he experienced what he believed to be a vision or visions from a supreme deity, whom he did not at first identify as God. These visions proclaimed to him that he was to rule the entire Roman world. So once things were settled in the north, Constantine invaded Italy itself in 312 C.E., driving on Rome in a series of battles and skirmishes to confront Maxentius. His brother-in-law had already successfully withstood two sieges of Rome mounted by Severus and Galerius, respectively the Caesar and Augustus in the east.

The final confrontation between the two one-time boyhood companions occurred just outside of Rome, but something extraordinary happened just prior to that fateful clash. A few days before the two sides met in battle, Constantine saw, or believed he saw, another vision. It was a cross in the sky with the Greek equivalent of the words *In Hoc Signo Vinces* inscribed under it, which translates from the Latin as "In This Sign Conquer." Then the night before the fight he had a dream in which God told him to use this sign to lead him in the coming battle.

Early the next morning he ordered the shields of all his men daubed with either a Latin cross or the anagram Chi-Rho. (History is unclear as to which of the two was actually used.) This monogram represented the name of Christ. Displaying this symbol for all to see, he proclaimed to his assembled troops this message from what he now identified as the Christian God.

Thus fortified and possibly also led by Constantine's labarum or military standard with the same symbol[37], they went into battle against Maxentius' forces. Maxentius in turn chose open battle to settle the issue even though he had fully provisioned the city of Rome to withstand a prolonged siege since he was a master of that type of warfare. He opted for a winner take all one day fight instead because the oracles and auguries were more than favorable. The day of battle, October 28, was the anniversary of his ascension to the throne, a most auspicious day, and because the prophetic Sibylline Books proclaimed that an enemy of Rome would die that day. Of course church history attributes this fatal decision to the influence of God.

The fighting occurred on the north side of the Tiber River in what is known to history as the Battle of the Milvian Bridge, which was a rickety pontoon structure spanning the river. With their backs to the Tiber, Maxentius' legions could not maneuver and were driven into its swirling flow. Maxentius himself drowned trying to retreat across the bridge. Legend would have it that Constantine was outnumbered by Maxentius, but the truth probably is that Constantine had the larger force.

In any event, Constantine thereafter became at least nominally a Christian convert. He went on to promulgate the Edict of Milan in 313 C.E. that provided for complete tolerance of all religions, especially Christianity. This edict was revised in part in 315 C.E. when he forbade Jews from proselytizing.

After a series of political maneuvers and battles, Constantine ultimately confronted his then co-Augustus, Licinius who had become the ruler of the eastern half of the Empire. They met in the climactic battle of Chrysopolis in 324 C.E., where once again Constantine triumphed.

With this victory Constantine became the undisputed ruler of the entire Roman world. In this capacity he was free to in effect make Christianity its officially protected religion, which he did. The rest, as they say is history. Christianity under Constantine and later his sons morphed through a series of steps and controversies into the Empire's, not only quasi-official, but also majority religion. And what evolved into the Catholic Church's hierarchal structure was fixed in place and centered in Rome.

In fact, Constantine considered himself to be the presiding bishop of this new state church. In that guise, in 325 C.E. he convoked the Council of Nicaea with 250 bishops from all across the Empire in attendance. Out of that council came what we know today as the Nicene Creed and so much more of what is now Christian dogma and belief.[38]

He also abolished crucifixion and made the cross, under whose sign he had conquered, the new iconic symbol of Christianity. Prior to his actions, Christians of the early church had seldom used it or the Chi-Rho monogram as such.[39] Since they hadn't, the question naturally arises as to why, if not inspired by God's silent voice, Constantine would have chosen the cross (what the entire Roman world understood to be the ultimate instrument of the Empire's most brutal form of humiliation, torture, and execution) to inspire his troops? It certainly wasn't a threat to Maxentius and his legions. As Roman citizens they were exempt from being executed in this manner. Under the laws of Rome only non-citizens could be hung from a cross.

As you think about this story, if you believe in God, it would clearly seem that God used Constantine and his armed clashes to take Christianity to its next level in evolution. As an individual, Constantine was no Jesus or Paul. He was instead a brutal and wily general and politician. Nonetheless, when needed, he both heard and heeded God silent voice.

It was Constantine's bloody battles and adroit political skills that gave him and God the Roman Empire and allowed them to make Christianity dominant within that realm. God used these brutal processes in order to ensure that Jesus' message of salvation, love, and forgiveness became the norm in the Rome world so that it could be handed down to us today.

To put all this into perspective, we need to ask ourselves what would have happened to God's message if Constantine had been defeated at the Milvian Bridge, or if, having won, he had not become at least nominally a Christian convert. In the eventuality of either of those two events, Christianity and its adherents would likely have remained a relatively minor segment of the Empire's population. Then, when the western half of the Empire collapsed a hundred or so years later under the pagan onslaught of the barbarian masses pouring across the Empire's northern borders, could Christianity have survived Rome's fall? Would God's message of love have been brought down to us today? The odds are that probably that message would not have made it but, instead, would have been lost.

The logical conclusion to be drawn from all this is that God desires positive change through evolutionary growth in both the natural world and in the condition of *Homo sapiens*. Humanity's God-induced growth includes not only physical growth, but intellectual, moral, and spiritual growth as well. In human terms, to achieve all this, God seeks agents of change both human and otherwise. And when God identifies these agents, be they individuals or groups, God, using a multitude of means – including silent suggestion, voluntary election, identified self-interest, compulsion, and, when necessary, violent persuasion, induces them to tackle the efforts needed to effectuate that change.

What God has made self-evident through both the history of the universe, and of humanity itself, is that God is not prepared to allow mankind's desires for the perpetuation of the status quo to stagnate humanity's advance toward God's goals for us. We humans can, in turn, use our free will to either elect to partner with God in these

endeavors or oppose God's efforts. It is our choice, but when we refuse to cooperate, or actively oppose God's purpose and plan, God
will find the means to push us back on track, even if that pushing induces incredible human pain and suffering.

WORLD WAR II

In more recent times God has demonstrated this point through the lessons of World War II and its aftermath. Despite nearly two thousand
years of ongoing instruction since God sent us the precepts conveyed
by Jesus and his predecessors and successors, by the first half
of the twentieth century the world was still struggling with the concepts of how to love, forgive, and sacrifice for ones neighbors and
others.

At the end of World War I, the embittered, but victorious Allied
Powers imposed extremely harsh, punitive sanctions on Germany.
These included the then staggering sum of thirty-three billion dollars
in war reparations, reduction in Germany's territorial size, and loss of
its colonies plus other sanctions including huge restrictions on the size
of its military. Coupled with a subsequent worldwide depression, this
nearly broke Germany as a nation. Out of their despair and angry re-
sentment, as well as fear of a rising tide of communism within their
country, Germans turned to Adolf Hitler and his German National Socialism. They did so because he promised the type of change they
wanted: a restoration of their pride, economy, and place in the world.
Unfortunately under Hitler these all turned out to be negative in nature;
nonetheless, through relatively free elections in 1932, Hitler came to
power as Germany's chancellor, the political head of its government.

Hitler, a megalomaniac warped by his own hatred and resentments, rapidly consolidated his and his party's power by destroying
the feeble German Republic and all opposition to his personal rule. In
doing so he was, in effect, using his perverted free will to oppose
God's teachings as well as God's plan for humanity's growth.

As they listened to him, the majority of the German people, buried in their own fears, anger, economic distress, and resentments, could not hear God's silently voiced messages of love and forgiveness. Most of them chose to block out those subliminal thoughts and God. Those that did not were either cowed or eliminated by Hitler and his henchmen.

Starting in 1933, Hitler proposed that the Reichstag, Germany's parliament, grant him the power to rule by decree. The principal political opposition to this legislation came from Germany's Social Democratic Party. After ninety-four of its members voted against Hitler's demand, twenty-four of their number were quickly assassinated and their party formally banned. This was followed that same year by the outlawing of all political parties other than the Nazis. Hitler's next step was to round up almost everyone else who openly opposed him. These people were either exiled, imprisoned, or murdered.

Thereafter, what little resistance remaining against Nazis rule was principally based in Germany's Catholic and Protestant churches. They at least heeded God's silent voice, but they were not enough. Their opposition was in response to Hitler's suppression of religious freedom. Several thousand Catholic priests, nuns, and lay leaders were either assassinated, or arrested and thrown into concentration camps. The Protestant denominations were similarly harried by the Gestapo and Hitler's police. Under the Nazi dictates the Bible's message and Jesus' teachings were perverted "to conform entirely with the demands of National Socialism." This was all done under Hitler's banner of "One People, One Reich, One Faith."[40]

The struggling opposition Protestants grouped themselves into what became known as the "Confessional Church" led by the Reverend Martin Neimoller. He was an unlikely candidate for this role. During WWI he had served in the German submarine service as a highly decorated U-boat commander. At first Neimoller, an ardent German nationalist, welcomed and praised Hitler's rise. But as he witnessed what the Nazis were doing to Germany, he swung toward opposing them. Thereafter Neimoller openly denounced their policies including their raging anti-Semitism. Ultimately, close to two thousand of his group's Protestant ministers, including Neimoller himself, were summarily arrested and herded into concentration camps where many of them died or were executed.

In their place, Hitler and his minions set up a "Reich Church" whose pastors had to take an oath of allegiance to Hitler personally. Following Hitler's lead, the Nazis viewed Christianity and its messages of meekness, love, and forgiveness as a religion teaching weakness, which in turn produced weaklings. Their new "church" proposed the eradication of the old Christian faith, replacing the Bible with Hitler's book, *Mein Kampf*, and the crucifix with the Nazi Swastika, a deformed and broken cross. They proclaimed a new "Christianity" centered on the individual figure and personality of Hitler.[41]

Its creed, as announced in 1937 by the Reich Church's head, was:

"The party stands on the basis of Positive Christianity, and Positive Christianity *is* National Socialism....National Socialism is the doing of God's will.... God's will reveals itself in German blood.... [Church leaders] have tried to make clear to me that Christianity consists in faith in Christ as the Son of God. That makes me laugh.... No, Christianity is not dependent upon the Apostle's Creed....True Christianity is represented by the party, and the German people are now called by the party and especially by the Fuhrer [Hitler] to a real Christianity....The Fuhrer is the herald of a new revelation."[42]

With this, a subservient Germany and much of its remaining clergy were almost totally obedient. Hitler was named president after a plebiscite in which 38,600,000 Germans voted to ratify his assumption of absolute power. And with this forced mandate in hand, Hitler unleashed his planned external onslaught on the rest of Europe, coupled with his internal extermination of the Jews and others he deemed "undesirables." This was all part of his nightmarish dream of racial purity for Germans and Germany.

Under his lead, the Nazis rapidly rebuilt Germany's army and air force previously frozen by the Treaty of Versailles at 100,000 men after the close of WWI. Hitler repudiated those limits, while the rest of the world sat by and watched Germany rearming at breakneck speed.

His next step in 1936 was to again flaunt that Treaty when he brazenly marched his new but still weakly armed forces westward across the Rhine bridges into the Rhineland. Although part of Germany, the Versailles agreements required that swath of land to remain a demilitarized buffer zone along the French and Belgium borders west of the Rhine River.

In the face of this blatant provocation, those two countries, the League of Nations, and the rest of the western world ignored God's silently voiced urgings to resist. Instead, in an exercise of their collective free will, they did nothing in response, other than to wring their hands in anguished concern. God's silent voice had not galvanized them into any sort of concrete action, even though at that time the French Army alone could probably have called Hitler's bluff and thrown the Nazi troops back across the Rhine.

With this failure to confront an emboldened Hitler, his next step was the more or less forced Anschluss union of Austria (Hitler's homeland) with Germany in 1938. Again this produced nothing more than protests from Britain and France, whose complaints were joined in by other European countries along with still greater hand-wringing. All the European powers stood around hoping someone else would take the lead and do something.

From there Hitler turned on Czechoslovakia. He wanted its provinces known as the Sudetenland along the Austrian and German borders with their ethnically and linguistically German populations. This time France and Britain were bound by treaties to support the Czechs, but instead of honoring those obligations, they again knuckled under to Germany. British Prime Minister Neville Chamberlain handled the negotiations with Hitler. In return for a worthless Anglo-German peace declaration, the British agreed to the partitioning of Czechoslovakia. Amazingly, the Czechs had no say in the matter. On September 30, 1938, Chamberlin made the hollow proclamation that, "I believe it is peace for our time."

How wrong he was. Hitler quickly consumed what remained of the Czech Republic and then turned his sights on Poland. On September 1, 1939, Hitler sent the troops of Germany's Wehrmacht pouring across the Polish frontiers with Germany. In the face of this act, Britain and France were no longer able to delude themselves and declared war against the Nazis. Now a truly world conflict had begun.

But it was a new kind of war with a German juggernaut known as blitzkrieg warfare as its centerpiece. Under its ferociously rapid onslaught, the Polish forces were annihilated while the French and British sat on the defensive behind France's wall of fortifications known as the Maginot Line.

After a brief pause to resupply over the 1939-1940 winter, the Nazis turned their attentions to these static defenses. With that France and the low countries of Belgium, Luxembourg, and Holland likewise fell to Germany's attacks. The British Army barely made its escape back across the English Channel from the beaches of Dunkirk in an epic evacuation many looked on as a miracle. Hitler's failure to wipe them out when he had them pinned with their backs against the Channel was the first of his great series of blunders.

The subsequent fall of the French nation and its democracy in 1940 along with those of the Low Countries were quickly followed by the democracies of Denmark, Norway, and the rest of the Balkan states. Greece held out a little longer, first against Mussolini's Italian invasion of 1940, so in 1941 Hitler launched an operation to bailout his fascist ally. With this, Greece was also overrun by the combined fascist forces, driving out the British Army that had been there fighting in support of the Greeks.

By mid-1941 Hitler and his Axis allies completely dominated Europe from the borders of Soviet Russia all the way west to the Atlantic Ocean. The only exceptions were a few small, neutral countries cringing in Hitler's shadow and a besieged Great Britain with just the English Channel standing between it and invasion. The prevailing public sentiment in the United States during all this was one of continued isolationism fueled by a feeling of safety behind the protective barriers of the Atlantic and Pacific Oceans.

So if we are inclined to look for the possibility of a loving God who interacts with humanity in the real world in order to channel us along the paths God desires, we have to ask what God did in these circumstances. When people, used their free will to ignore God's subliminal urgings to resist and confront the fledgling monstrosity that became Nazi Germany, a Germany whose emboldened leaders moved from one brutally negative and animalistic atrocity to the next even worse atrocity without seeming end, how did God respond?

Arguably God responded in two ways. First, God engineered events that both shocked and galvanized into action those who had the capacity to take on the nihilistic forces of the Axis powers. And secondly, God employed God's silent voice to misdirect those very nihilists into the blunders that would lead to their downfall. In short, God silently and subliminally urged Hitler and his lackeys as well as the Japanese to take a series of incredibly imprudent steps. These were acts those men dreamed about and desired in order to gratify their insatiable appetites for power; they were steps and actions their more sensible military advisors urged and pleaded with them not to undertake.

How did God do all this? First, confronted with the ultimate free will of human monster, Hitler, God pitted another equally monstrous man, Joseph Stalin, and his equally atheistic Union of Soviet Socialist Republics against the Nazi threat.

To create that conflict it would seem God subliminally urged Hitler to do just what Hitler always wanted to do. That is attacking Stalin and the USSR in order to gain the vast Russian land expanses to

Germany's east for Hitler's dreamed of Aryan race and the "living space" he thought it needed. Hitler planned to use those lands for the expansion of a Greater Germany and what he saw as an every growing pure Germanic population. God played on Hitler's hubris to coax him into this blunder of all blunders, Germany's 1941 invasion of Russia.

The best generals of the German Army and its General Staff feared that a two-front war in both the east and west that such an invasion would produce would be a fatal mistake for Germany. History from Napoleon on to the lessons of World War I screamed this message at them. As a result, opposition to a two-front conflict had become enshrined in German military thinking. Plus, even though Stalin trusted no one, especially not Hitler, the Nazis had successfully inked a non-aggression pact with the USSR in 1939. And this pact secured Germany's eastern borders. Under its umbrella the two despotic regimes had divvied up Poland and the Russians were supplying Germany with critical raw materials needed to support the Nazi war machine.

Hitler, with God silently playing on his ego, ignored these advantages and brushed aside his generals' objections. On June 22, 1941, he launched Operation Barbarossa, a tidal wave of 3 million men made up of 116 infantry, 19 Panzer and 9 line-of-communications divisions. Buoyed by Germany's series of lightening successes in Europe, he expected to conquer the USSR in no more than seventeen weeks - well before winter weather set in.

Instead, the war on what became the Eastern Front was to last almost four bitterly long years including four brutal Russian winters and consume in excess of 200 of the Nazis' best army divisions as well as the better part of the Luftwaffe, Germany's air force, along with millions of its fighting men. And it led to the 1945 fall of Nazi Germany after Hitler's suicide deep in a Berlin bunker under the bombed out rubble of the German capital as it was being overrun by the Red Army.

In this "no quarter asked for or given" fighting, the Soviets lost over twenty million lives including more than eight million members of the Red armed forces. Only a ruthlessly barbaric dictatorship such as that crafted by Stalin could have forced such horrific sacrifices on a people. It was a society in which you had a choice of either fighting the Germans to the death or being shot by your own side if you did not. So the Soviets fought, as did their German enemies for the same reason. The two sides lost a million men between them in the single pivotal 1942-1943 struggle for the city of Stalingrad and its control of the Volga River.

While all this was going on the United States was jolted out of its isolationist mindset by another brutal fascist dictatorship, Imperial Japan and its sneak attack on the United States naval base at Pearl Harbor, Hawaii, on Sunday morning, December 7, 1941. Japan's leadership launched this strike because it felt its war machine was being strangled by the US embargo on the oil and scrap metal sales Japan needed to keep its military conquest of China running. The Japanese thinking went that if it crippled the US Navy in the Pacific it would have a free hand in taking over the Dutch East Indies oil fields and the other vital raw material sources in Southeast Asia they desperately needed. Japan's attack, however, was a blunder equal to Hitler's invasion of the USSR.

It brought a previously pacifistic, but now enraged, United States into the war in both the Pacific and Europe. America cannot claim it won that war alone. As a nation it lost approximately 416,000 military dead compared to the Soviets' eight million. But the US became the arsenal that provided crucial war supplies to both Britain and the USSR. Along with its ally Great Britain it fought and won pivotal battles in North Africa, Western Europe, the Pacific, and the Atlantic. The most notable of these included the Normandy invasion and the Battle of the Bulge. The US and Britain knocked Italy out of the war, leaving Germany to fight off the Allied advance up that peninsula with men it couldn't spare. All of this kept a critical mass of Germany's fighting strength tied up on the fatal Western Front of the two-front war the Nazi military dreaded. This was just the disaster the Wehrmacht's generals had most feared and Hitler, in his God-induced hubris, had so blithely disparaged.

To put this in prospective, we must look at the alternatives and consider what the outcome of WWII would have been if, say, during WWI the Kerensky provisional government of Russia had not been overthrown by Lenin's Bolsheviks and if Stalin's Soviet regime had never come into existence. If instead, Kerensky and his successors had tottered forward as a more or less democratic government into what became WWII.

It can only be assumed that such a Russian democracy would have been far weaker than what became the USSR. The odds are that it would have crumbled in the face of a Nazi onslaught, just as so many other European democracies had when they suffered German invasion. Could any democracy have sustained a loss of nearly fourteen percent of its total population, and have another thirty percent either wounded, starving, or both, without collapsing?

If this had happened or had Hitler listened to the sound advice of his generals and not invaded Russia until he'd overcome Great Britain in the west, Germany's 200 divisions, which were ultimately consumed in the Eastern Front fighting with Stalin, would have been available to defend the Nazis' Atlantic Wall and confront Britain in North Africa. There would have been no British defeat of Rommel and his German Afrika Korps in the fall of 1942 at El Alamein, and the Suez Canal, Egypt, and the Middle East's oil fields would probably have fallen to the Germans. There would have been no successful Normandy invasion. Hitler would have been secure in his conquered Europe and have had the time to develop Germany's own atomic bomb that his scientists were working on. And the extermination of European Jewry would have been completed. The world would have become a far different place today.

But that's not what happened; instead, a previously sure-footed Hitler made blunder after blunder on both Germany's Eastern and Western Fronts. The United States was forced into war with Japan's attack. Hitler then obliged America to also fight Germany by declaring war on the US in support of his nominal Japanese ally when he could have ducked that fight.

He did so because of two things. By December 1941 he was waking up to the fact that the Soviets were not the easy pickings he had expected. So he thought he might need Japan's help via a Japanese attack on the USSR's far eastern defenses a la their victory in the Japanese Russian War of 1902. And Hitler held America, Americans, and the US soldier in contempt.[43] All this led to the total and unconditional defeats of Italy, Germany, and Japan. God's silent, subliminally voiced urgings told Hitler and the Japanese leadership what they wanted to hear and believe.

While no one can prove God goaded Hitler and the other Axis leaders into all these fatal mistakes, it is not unreasonable to perceive the possibility of God's hand and silent urgings as the agents that egged on such a sequence of otherwise irrational, hubris-based events. They produced the outcomes God required to advance God's loving plan for humanity, even though they entailed enormous human suffering and incredible amounts of violent conflict when the western democracies initially refused to act early on.

THE COLD WAR AFFTERMATH

Contrary to its historic leanings, the United States did not retreat into comfortable isolationism after WWII. Nor did it and its western allies impose crushing World War I type war reparations on their former enemies. Instead, while the forces of the USSR looted the eastern half of Germany and annexed Japan's northern most islands, America led the effort to rebuild its former enemies and the rest of Europe through the Marshall Plan. This was to be a bulwark against the Soviet war machine. And as part of occupying Japan and Germany, the US and its allies guided those two counties toward the democratic and non-militaristic processes that now govern what have become those two vibrant nations today. Thus it would appear that God's lessons of forgiveness and love had finally come to the point where they could be learned on national levels.

The Soviet Union, whose communist economic regime and huge military outlays were unsustainable under such a system, finally collapsed in the 1990s. Its demise ended, at least temporarily, the Cold War with the West. In comparison, Hitler's evil system of German National Socialism, if left undefeated and free to exploit all of Europe, could have survived far into the future.

But before the Soviet collapse two extraordinary things happened. The first of these was the Cuban Missile Crisis of 1962. This was at the height of the Cold War standoff between East and West, between capitalism and communism. And it was as close as the world has yet come to all-out nuclear war.

The crisis began with America's stationing of strategic ballistic missiles on Turkey's northern borders aimed at the adjacent heartlands of the Soviet Union. The Russians responded by attempting to covertly introduce a large force of their own nuclear armed missiles into Castro's Cuba. Once in place and aimed at the United States this *fait accompli* would be disclosed to the US as the counter to the American ploy in Turkey, but American intelligence got wind of the Soviet efforts.

After the now famous U2 flights and other aerial photo reconnaissance confirmed the presence of the first Russian ballistic missiles and missile launchers being set up on the island, the US instituted a naval blockade of Cuba to halt any further missile arrivals. With that the world was faced with a crisis of potentially cataclysmic proportions as the US and the USSR edged ever closer to a full-fledged nuclear exchange that would have produced a holocaust of death and destruction.

It fell to two men, United States President, John F. Kennedy and USSR First Secretary, Nikita Khrushchev to find the path that allowed both sides to back away from this dangerous confrontation without either side losing too much face. Each man's respective militaries and much of both countries' top leadership were urging them to stand up to their enemy and not blink.

Ultimately, though, those two individuals led the efforts needed to forge a compromise in which the Soviet's first withdrew their missiles from Cuba and turned around the Russian ships testing the blockade while the Americans thereafter removed their ballistic missiles from Turkey. This allowed both sides to avoid what could have become nuclear Armageddon.

The second event was Mikhail Gorbachev's rise to power in the USSR in the mid-1980s. After becoming general secretary and head of the Soviet government, he didn't challenge the dismantling of the Soviet block of Eastern European countries and the fall of the Berlin Wall by sending in Soviet tanks as had happened in Hungary in 1956 and Czechoslovakia in 1968. He sanctioned the USSR's military withdrawal from Afghanistan in 1989 and ultimately allowed the Soviet Union itself to, in effect, go out of existence without interposing armed force to prevent that happening.[44]

This transitional change from communism to something approaching capitalism was extremely hard on many of the people who had to live through it. Even today large numbers of them are still struggling to cope with the negative impacts on their individual lives produced by this traumatic shift. Many of them were thrown into poverty as a result, but it was all done without major armed conflict.

We must certainly consider the possibility that it was God's silent voice that successfully urged all three men to find the necessary ways to avoid war over the events confronting them. They did so even though each was faced with incredibly powerful forces within their own governments and societies who opposed changes to the ways they and their countries had previously responded to such threats and challenges. All three men paid heavy prices for their leadership. Two years after the missile crisis Khrushchev was deposed by the rest of the Soviet hierarchy and Kennedy was assassinated. In Gorbachev's case, his actions were done at the cost of his own personal standing with the Russian people. The fact that the three listened to God meant that, despite the negative impacts all this produced on many people, God was not driven to the extreme violence we humans have often forced God to employ when we do not listen to God's subliminal urgings.

It is now clear that key events in the spread of God's messages of salvation, forgiveness, and love were crystalized in unique ways for humanity through violence such as Christ's crucifixion and the subsequent outward spread of what became Christianity from Israel to the surrounding gentile world in the face of physically violent opposition. Arguably, all of this was engineered by God, just as Buddhism and Islam were spread, as part of God's purpose and plan for the advancement of humanity's growth out of our past animalistic imperfection toward a greater moral and spiritual perfection. And most recently those messages survived one of the greatest threats of extinction they ever faced in the forms of atheistic Nazi Germany and an equally atheistic Soviet Russia. Both Hitler and Stalin would have gladly wiped out Christianity and its message had either of them ultimately triumphed. God utilized their countervailing ruthless dispositions to negate that threat.

In defeating these two challenges to God's message and plan, God would certainly have had no doubt as to what the costs were going to be in terms of the horrendous human toll of terrible death and suffering that would accompany the process of overcoming these menaces, yet God went forward with it nonetheless. In this same way, almost two thousand years earlier God was fully cognizant of the agonizing suffering the Romans were going to inflict on Jesus and his followers as part of the spread of Christianity and went forward with it as well.

If we are willing to acknowledge the possibility of God's providential influence on this trail of events – starting with the creation of our universe and moving on to the birth, propagation, and protection of humanity itself, as well as Christianity, plus the rest of our major religions, and modern civilization – we must be prepared to take the next step in our intellectual analysis of God's methods. That step is to recognize that right up to today our loving God is still prepared to take whatever courses of action are necessary. This includes the use of human suffering, even if it requires suffering on a worldwide scale, when we humans offer God no other alternatives to achieve God's ongoing plan.

There are important lessons for us in all of this as we attempt to gain a clearer understanding of God and God's ways in dealing with the free will choices we make in response to God's challenges: challenges presented to each of us individually and collectively.

Chapter Fourteen

FREE WILL

We have already mentioned free will any number of times as we've looked at how God does what God does. But for our purposes, free will is only of interest when we talk and think about the human condition. It has little or no impact in the Earth's greater biological world outside of what we humans do or don't do.

Setting humanity aside, the animal conduct found in the rest of that world is, as we've previously discussed, driven by the Darwinian/evolutionary imperative of eat or be eaten. And that imperative has no room for notions of right or wrong, good or evil.

But when it comes to individual humans, God seems to have gone to great pains to grow not only our intellectual powers, but also our capacity for moral, ethical, and spiritual decision making. The ability to separate good from evil in all their gradations is now an inherent part of our thinking. Then to act on an understanding of these concepts in terms of our spirituality, ethics, and morality is basic to our relationship with God, as well as each other. And the critical factor in the growth of these attributes is the human free will God has so carefully nurtured in each of us and seems to prize.

But what is free will, what is its function, what does it do? First, human free will is our potential for unfettered independent choices as to what we think, believe, do, and say within the frameworks of the time and cultures we live in. It enables us to craft each of our own individual lives. Then collectively, it permits us to participate in shaping our societies, environment, and the various religious belief systems we elect to adhere to or reject. It likewise allows each individual to choose the political and social organizations we want to live under, and it dictates how we respond to one another in ever widening circles of relationships, starting with our one-on-one interactions, and then expanding out from there all the way to humanity as a whole.

Free will gives us the freedom to be selfish and self-centered or altruistic and caring. It enables us to elect who and what we care about. Our free will decisions empower us to choose to be conservative, moderate, liberal – or some complex amalgam of all of these – in our thinking, beliefs, and associations. It permits each of us to make layer upon layer of free will choices over our lifetimes that end up creating our personas as human beings. Its exercise is a never-ending lifelong process.

Our individual and collective free will are, at one and the same time, both powerful and fragile. Based on their exercise we choose or seek to choose what we do as our life's work, where we live and play, with whom we associate and who we select as life partners.

On the other hand, our free will are often compromised or over-whelmed. For ill or good, other people, our families, societies, and governments all seek to limit and channel the exercise of our individual and collective free will. These forces try to mold the outcomes of that exercise, and with that molding, control as well, what we do, say, and think. In short, free will and its exercise are at the core of who and what we are. It goes a long way towards dictating our lives as human beings.

In the midst of all this, our free will can also be overpowered by various psychoses, agents, and forces starting with our addictions. In terms of addictions, these may be as obvious as alcohol, drugs, sex, money, or power. They may be more subtle ones like tobacco, caffeine, overeating, or the need for approval or group inclusion. Once addicted, though, it is incredibly difficult to invoke our free will to break away from whatever our addictions are.

Other factors, like fear, can likewise defeat our free will. When faced with physical danger or social pressures, we may know what we ought to have the courage to do, but often we are too intimidated or fearful to muster our free will to do it. Inertia can, in the same way, inhibit the positive use of our free will. We find it much easier to let negative free will take over when we are confronted with the necessity for action. All too many of us stubbornly resist needed change when that change requires confronting threats, sacrifice, or discomfort; giving up of our self-indulgencies and ease; or abandoning our self-interests.

Nonetheless, free will and its exercise are essential to who and what each of us is on an individual basis. Collectively its use or suppression defines and shapes all of our societies. In that setting, free will also allows us to decide when and to what extent we listen to God's silent, subliminal voice. Its exercise permits us to block, ignore, or reject that voice. In fact, it controls our personal interactions with God. And apparently God wants it that way.

But why is that so? Why is God so patiently tolerant of our halting advances out of our animalistic and selfish heritage? Why does God allow us to exercise our free will to harm one another, our earthly habitat, our environment, and relationship with God – all the while ignoring God's teachings?

A significant number of those who deny God's existence base their denial, in part, on the fact that humanity is so flawed and imperfect in its exercise of free will, when God, if God existed, could have simply commanded it otherwise. For them, if God were real and not just a figment of our imaginations, that God would have made us far more nearly perfect from the beginning.

Responses to this position that come to mind are twofold. First, as we've already pointed out, God is not our servant whose primary function is to make us and the things we want come out the way we want them to. Second, the fulfilment of God's plan for humankind seems to require humanity in its individual and collective components to have had the living experience of facing and overcoming our imperfections and flaws in the exercise of our free will choices. Simply being perfect from the start, while intellectually or academically recognizing the possibilities and potential impact of failed or negative exercises of free will, is not enough. That would, by definition, create the lack of a critical element in the totality of experience needed to allow us to approach a future perfection our loving God seems to be guiding us toward.

Thus, in much of God's interactions with us, God must rely on influencing our thoughts and conduct through that silent voice that so many of us think of as our conscience – something we have already said so much about. In this context God is apparently prepared to tolerate our imperfections and free will denial of God's directions.

Therefore, we in turn have to be willing to accept that same fact and its impact on us. As we confront this truth we need to also recognize its corollary: that our journey towards God's loving and perfect future is just at its beginning stages, still has a long way to go in terms of evolutionary time, and doesn't travel in a straight line.

In this light it would appear that God's focus is, generation by generation, on the growth of humanity's capacity for exercising our free will in a positive fashion. This maturation is occurring in conjunction with our growing intellects and our expanding moral, ethical, and spiritual capacities.

The recognition of God's patiently loving efforts in this regard only leads, however, to other questions. The first of these is why does God require us to grow out of such imperfection toward a far distant perfection? What purpose does this requirement serve in terms of God's plan?

As we parse out the direction God is pointing humans towards by fostering free will, will it be enough for us to simply reach a level where we can freely return God's unconditional love equally unconditionally? At that future juncture will we have achieved the ultimate in coexisting with God for eternity in some sort of a blissful state of cohabitation? And how will such an eternal reciprocity of love serve God's plan? Or, alternatively, will God require something more of us, and if so, what?

If we assume and believe that plan includes a non-corporal yet sentient existence for humanity subsequent to our earthly lives, God's focus on growing our capacities and strength for the free will exercise of moral and ethical decision making may have a number of additional implications beyond simple reciprocal love. The first of these is that in whatever that heavenly afterlife is, it will require us to be able to exercise our free will in such a setting. We may have to make choices based on our free will that include deciding whether or not to follow God's voiced desires and directions. The further implication is that those directions may entail some form of ongoing, but needed, sacrifice that on a purely selfish plane we might rather not make. We may also have to elect how we deal and interact with not only God, but also all those other souls that populate such a heaven as well. If this is true, it would seem that God is preparing us in both this life and whatever comes after to grow in positive ways that will advance God's overall plan no matter how far we have to stretch in doing so.

Embedded in all this is yet another question. In considering the possibility of transitioning the use of our free will from our corporal earthly existences into a non-corporal heavenly realm, we are faced with an obvious hurdle. Here on Earth God has largely refrained from speaking to us in other than subliminal, silent tones. God has let those tones percolate up out of our subconscious. We often only recognize these communications as our consciences speaking to us.

But what happens in heaven? As we've discussed, in this existence when God has spoken directly and more or less audibly to any given individual that person's free will has been overwhelmed, so assuming that in a heavenly existence we will be directly in the presence of God, how does God allow us to retain the use of our free will without becoming slavishly and unthinkingly addicted to an adherence to God's directly experienced voice? Does God continue to communicate to our non-corporal souls in some form of similarly subliminal processes, or will we have developed a sort of non-corporal filter that allows us to hear God while still retaining those free will? Without our bodies we will no longer be at the mercy of our physiological addictions. Will the same hold true for our intellectual, psychological, and spiritual predilections as well? Alternatively, will Heaven consist of various levels that will in some way separate the levels of our interaction with God, thus protecting each individual's free will capacity to act?

All this in turn implies that whatever that heavenly existence is, it will be far more complex and challenging than most of us currently conceive of it. Instead of an eternity of blissful ease where we simply bask in God's glory, we may well have to continue our courses of positive growth on a free will basis. If that growth doesn't continue we need to be prepared for the possibility of negative consequences as a result of our failure to advance.

This possibility, in its turn, brings up the questions of what will be the uses a loving God, who is not our servant, have for each of us in that far distant future once we achieve the perfection God is directing us toward? Why does God want us to share such a perfect eternity in God's presence? And in such a setting what are the challenges that could be presented to us by the uses God has for each of us there?

Chapter Fifteen

DEATH

Before we can even hope to get to such a heavenly existence we have to deal with death. Most view death as the ultimate change and challenge in our earthly lives. No matter how hard we try or how much we would like to, it cannot be avoided. It looms in front of each and every life that has ever been lived. Many of us rail at it. Others seek to ignore it. Some hide from it. Most fear it in one way or another. But it is ever present and one of the central facts all of us have to deal with as best we can.

Why? Because death is critical to God's plan for all biological life. In evolutionary terms it makes room for each succeeding generation. It allows every form of life to have its crack at evolutionary growth, thereby improving or winnowing out species after species over the time frames of evolution. Without death, life's breeding processes across all the living species would either slow to a crawl or overwhelm creation. Either way the total numbers of living entities competing for space in which to exist here on Earth would so multiply as to make life not worth living. Death's absence would effectively stunt and deform all growth while at the same time inhibiting the evolutionary advancement God requires.

Therefore, God's plan needs and requires biological death generated by the shortcomings and degenerative failure of all the cellular structures that make up organic life. This is the first form of death God uses for God's plan. It is the way most of us will personally encounter death.

However, God uses death for far more than as just an evolutionary broom to tidy up the biological world. This is especially true in terms of God's growth of humanity.

So let's explore how, having instituted death as part of the biological world, God achieves God's loving ends at least in part through the mechanism of death. We need to do so because death's knell acts as the goad for so many aspects of our human lives. It all starts with God's recognition of our inherited animal instinct to perpetuate our individual genes through procreation, and our related love for our offspring that are the product of that procreative drive.

With that love as a base, God has subliminally encouraged us to become aware of, and care about, those we encounter surrounding our nuclear families. Recognizing other people's myriad of physical distresses, and our common experiences in the face of death, God has silently urged us to seek to alleviate the suffering of others. This extends to even trying to help them stave off death itself as long as possible and hoping they will help us as well.

These processes underscore the complexity and interdependence of all human lives and relationships. They teach us just how much we lack in self-sufficiency, as well as how critical we all are to each other. This then encourages us to value our relationships with the rest of the human community and ultimately with God. It also fosters our love and need for wider and wider circles of our fellow human beings and the world that sustains all of us.

Recognition of our own inevitable deaths impels most of us to care about the futures of those we will have to leave behind both individually and institutionally. It makes us think of, and provide for, them in our looming absence. We seek to plan for their ongoing existences, needs, success, and wellbeing. We do so in order to protect what and whom we care about and love, as God has taught us to love and care for others.

The existence of death also tends to make us consider what we are building in terms of the societies and world we are bequeathing to those who follow us in life's ever succeeding generational march. Will they have fond memories of us, honor our accomplishments, or even remember us? Or will they curse and revile us for what we did to them and the disasters we left behind? God has patiently been leading us to consider these questions in the face of our own approach to what we perceive as a one-way door at the end of our biological lives.

The second kind of death God asks us to deal with is the result of the dangers presented by the world we live in. As we've previously noted, it is a world that abounds in challenges and threats that can and do kill people in great numbers; be they famine, lack of or bad water, hurricanes, floods, tornadoes, blizzards, cold, heat, earthquakes, tsunamis, plagues, disease, and on and on, each takes human lives.

God uses the menace each of these presents to life as another tool to force us to band together to deal with them out of love and the need for mutual protection. The pitiful sights presented in the aftermath of these disasters move more and more of us to a shared desire to help their victims no matter the color of their skin or their geographic location or culture. Pictures and videos of starving or maimed children can't help but move people into loving action.

Our responses, both emotional and intellectual, to these events shape our societies and us over the generations. In the face of God's silent prodding we find it harder and harder to simply turn a blind eye when they occur. We and our organizations – religious, charitable, or governmental – feel compelled to push back against such threats to life. All of this is part of God's ongoing effort to advance us out of our animalistic past toward the future God is directing us toward. These responses grow our capacity to love each other and God.

We have to confront death in other guises as well. It is not just created by natural, biological, and environmental forces. Death is also generated by human conduct and volition. Driven by the very animal instincts God is trying to wean us away from, we humans often use death for our own purposes. Out of hatred we kill to wreak vengeance on others. Out of lust for power we employ death as a tool of intimidation and control. We murder to settle disputes ranging from one-on-one confrontations, to global wars. Out of greed we slaughter or impoverish entire populations, leading to their untimely demises. We pedal products and drugs that we know kill their users. And countless times we kill or maim our fellows out of sheer thoughtless negligence.

God teaches us to respond to such human acts by first generating and then living by laws and mores that resist and inhibit this sort of conduct. Our earliest records show us God instructing humanity that such practices are both wrong and forbidden. We've already pointed out God's command to Noah after the flood, when supposedly there were only the eight members of Noah's family left following God's destruction of all the rest of human kind:

> "If anyone sheds the blood of man,
> By man shall his blood be shed;
> For in the image of God
> has man been made."
>
> (Genesis 9:6)

As likewise written in the Book of Exodus, we find among the Commandments God proclaimed to Moses on Mount Sinai the first of the "thou shall nots," which is the admonition "You shall not kill." (Exodus 20:13) These are just two of the many examples of God's steady drumbeat of loving instruction, designed to lead us away from our animal past of viewing the taking of human life by other humans as acceptable behavior.

God's constant effort has been, instead, to teach us to love and respect each other just as God loves and respects us. From this base has sprung the incredibly intricate webs of interlocking laws, covenants, customs, and practices that now begin to blanket the entire world, all focused on protecting life and inhibiting death inflicted on humans by their own kind. Unfortunately, these are lessons we are still far from mastering.

143

There is also another kind of death we should consider. It is one that we must look at in a different light. This is self-sacrificing death. God has taught us how, out of love, to be prepared to sacrifice ourselves and our lives for the protection of others. One of the foremost examples of this was and is Jesus' willingness to accept death on a cross for the good of the world. His unforgettable example of this form of God-taught love has reverberated down through two thousand years of time and has changed humanity. By riveting our attention on Jesus' sacrifice, God has also focused humanity on the lessons of love Jesus spent his ministry teaching to all those who would listen.

The willingness to risk one's life to not only save others, but to preserve and protect institutions, religions, and concepts – or to put ourselves at hazard in wars to defend them – has become a hallmark of positive human growth. It draws us past our selfish animal instincts, into a realm of love of others whom we may not even know. It expands the soul and leads us to be more like Jesus and less like self-centered brutes.

But no matter which form of death we consider, central to all of them is the termination of our earthly biological existences. The approach of this biological end leaves us confronting a factually unanswerable question. Is there some form of existence after our deaths? Some believe that there is nothing. They believe we simply cease to exist except as memories in the minds of other humans and in the written or otherwise recorded accounts of our lives, actions, and thoughts. But the adherents to this position must base their certainty on belief and a sort of faith since it is not provable as fact.

Others, and many of the religions they adhere to, believe to the contrary. For those in this category there is faith that there is some form of life that continues after our biological demise. For many this belief may take the form of the concept of reincarnation in later biological lives. For an even larger group it may be faith in the survival of their souls in another dimensional existence they think of as Heaven or the afterlife. There are any number of variants and variations on these themes, but they are all still based on faith and belief. No one can factually prove them to be true, or for that matter, false.

But why is that? If God exists, why hasn't our loving God at least seen fit to give us a factual, as opposed to a theological, basis upon which to once and for all definitively determine if our existences stop with biological death, or somehow continue on thereafter? Would that be too much to ask?

One answer to this riddle may be that such knowledge would not advance God's plan of growth for us. It could in fact be detrimental to God's efforts. If we knew for a factual certainty that our existences and that of all human life continued on after our physical deaths, would we be as concerned about the fates of those we leave behind? Would love play as great a part in our lives? Would we care as much about institutions, systems of governance, and religions that direct earthly societies if we knew without question that they were only part of a transitory phase of our existences? In fact, a relatively short-term interlude that was to be followed by an infinitely longer non-corporal, pain free, state of conscious being?

Would we be as moved by the deaths of distant peoples if we absolutely knew they were simply moving on to a far better plane of existence? Would we be as reluctant to kill other human beings if we had no doubt that we were just facilitating their transitions into a vastly happier non-corporal, eternal life? Would we try as hard to cure disease and alleviate pain?

The answer to all these questions is an emphatic no! The care and emotional cost involved in responding to others takes a lot of effort and sacrifice many of us would view as an unnecessary waste if we were certain of what was to come. That certainty of knowledge would have a major negative impact on human thinking, conduct, and relationships. It would detrimentally affect all of our societies' institutions and activities. Why expend the energy needed to build something in this short-term world, when an infinite existence awaits all of us? In the face of danger or difficulties many, if not most of us, would probably take the easy way out and give up, expecting that we and those around us would simply move on to the next plane, and it would inhibit our thirst for knowledge. We could just ask God all our questions when we got to heaven.

On the other hand, if God gave us absolute proof that there was nothing after our worldly existences, that as far as we were concerned everything we could ever experience came to a dark blank wall of non-being with our deaths, what would many of us do? How many would adopt the Roman gladiators' motto of "eat, drink and be merry, for tomorrow we shall die"? Would we try to grab all that we could in terms of wealth and comfort without concern for that attitude's impact on others or our world? How much more greed would there be in our human communities if there were no possibility of a future reward in a heavenly existence? Would we be as concerned about God's judgment of our conduct and lives in this world if there was absolutely no possibility of rewards or punishments in a hereafter? Would we be as willing to risk death and the permanent end of our experiential lives for what we believed in, or to save others?

The answer to these questions is that we would have to assume that in the face of such certain knowledge of ultimate nothingness, our human conduct, when spread across the entire race, would for many of us be far less charged with love, caring, and self-sacrifice. An attitude of me first and get what you can, because this is all that there is, would be far more likely to predominate.

In light of these two probable, but negatively parallel, results produced by such opposite certainties of outcome, it would seem that God has instead elected to employ the ambiguity of our not knowing for sure, one way or the other, what follows biological death. This is the course that best serves God's purpose and plan for humanity. In the place of certainty, God is lovingly, but indirectly, teaching us about the possibility of an eternal afterlife to be shared with God. God has employed the storytellers we have talked about to promise us such a life.

And God has used prophets and apostles, including Jesus' followers, as well as Jesus' himself to preach this message to us.

But with a universal confrontation with death in front of all of us, we are forced to accept or reject this promise based on our individual faiths, and not on fact. This is what best drives God's plan for the advancement of humanity. It is what produces the greatest positive growth toward our future human achievement of something approaching perfection. In this sense God is far more interested in advancing God's plan than in pleasing or mollifying us. God can answer our questions after we experience death if that serves God's purposes.

Chapter Sixteen

OUR SOULS AND HEAVEN

Let's assume the answer to the question we've just looked at, the "is there life after death" puzzle, is yes. And let's further assume that this life consists of an eternal or otherwise unmeasurably long-term, non-corporal existence in another dimensional realm. The next question then becomes how do we get there and in what form?

The simplistic answer to this is that our souls, which we've been talking about, are saved by God and transported to that heavenly dimension. But what is a soul? What does it consist of? Where does it reside within us? How does it function? And is it individually the essence of each of us?

Despite all of our science no one has ever physically isolated or even detected a soul. Thus, they can't be measured or empirically tested yet. We can't even prove that they exist. They are another one of those mysteries of life we have to take on faith, yet so many strongly believe in their existence.

Despite this lack of proof, we can make some interesting guesses about souls. In doing so we should start our discussion by once again looking at the ways God interacts with each of us. Based on our analysis to this point, it would seem that down through the ages God's efforts have been focused on growing each individual soul's capacity and capability to become other than animalistically self-centered and flawed. This is the ongoing struggle to shed such selfish proclivities and leave them behind us.

So whatever a soul is, it has to be able to evolve towards an ever greater capacity for caring about everyone else we interact with. In short, to meet God's requirements souls have to advance, generation by generation, toward achieving a more loving relationship with others and with God, as well as growing morally and spiritually. But this still doesn't suggest what our souls are in terms of what they are comprised of, how they function, or what their use is.

To be of any use to God, our souls must be something that actually exists, just as God must be real to be of help to us. Thus souls need to be far more than a concept; instead, they have to be something that, after our biological deaths, still physically retains our earthly experiences, understanding, and character. They have to have a capacity for learning and intelligent communication with other souls, as well as with God. They must be able to maintain all these abilities after separation from our physical bodies.

Since modern medical science has eliminated the possibility of souls being some form of solid biological matter contained within the human body, they must be non-corporal in nature. The logical conclusion to be drawn from this non-biological requirement is that our souls have to consist of something akin to cohesive, internally interconnected energy packages. These packages also have to be something that will not dissipate over time or with their transition from one dimension to another, no matter what that entails. And after that transition they have to retain at least the essence and positive aspects of our individual characters while, hopefully, shedding all of our negative animalistic traits. This also implies that these soul packets are recognizable as distinct and unique individuals that can be differentiated one from another as well as by each other and by God.

But what sort of energy are we talking about? If it is electrical in nature it should be detectable by modern science, just as we can now detect, measure, and trace the electrical impulses that flow through our individual brains and nervous systems. These, however, seem to cease with our deaths. When someone is brain dead there is no measurable electrical output from and through the higher levels of their brain functions, even though the physical bodies and other organs of the host may still be functioning for a time.

While a composite of all our brain's activities might be a candidate for our souls, that composite doesn't seem capable of sustaining itself without the biological structures and neural pathways its activities normally traverse within us. This is especially true since those functions are based on a form of electrical energy. Such a current requires a medium like nerve fiber, liquid, or cabling (fiber or metal) to travel along. If instead of being a current in nature, they more nearly resembled a wave akin to light or radio frequencies, they still would need an initial physical transmitter. In either event, both ultimately start to dissipate over time and distance; therefore, it wouldn't seem that these possibilities are viable candidates for what makes up our eternal souls.

Beyond these suppositions we begin dealing in the realms of pure conjecture. Modern physics, however, may point us in the direction of some fertile areas for our speculations: specifically the possibility that souls are made up of something akin to dark energy and dark matter, and modern science cannot directly detect either of these dark entities and has very little in the way of concrete ideas as to what their makeup is. Our scientists just don't know enough to be able to say what they are other than they have to be there. They can, however, infer their presence through their impact on the universe as a whole. Without these dark entities the visible universe, with all its intricate structures made up of an endless parade of galaxies and all they contain, simply wouldn't function or exist. And that includes us.

If you separate the two, they have been calculated to consist of seventy percent dark energy and twenty-five percent dark matter, making up ninety-five percent of everything in our universe. Dark matter seems to form an invisible skeletal framework upon which all the visible galaxies and galaxy clusters all across the universe are strung. The resultant totality is held together by some form of gravity. The dark energy portion of this equation is what is driving everything in the universe outward from its big bang origins at ever accelerating speeds. This effectively stretches and expands space itself in all directions.

Thus, we shouldn't be either surprised or discouraged by our not being able to scientifically identify human souls when we still can't directly detect what makes up ninety-five percent of our universe, even though we know it's there. And by the same token, just as we can observe the effects of dark matter and dark energy on our universe without directly detecting them, we can likewise indirectly perceive the impact of our undetectable souls on individual human beings as well as their collective effect on all humanity.

Therefore if our souls are similarly made up of something akin to packets of dark matter and dark energy, we wouldn't be able to detect them, just as we can't detect those dark entities even though they have to pervade our entire universe. Now, moving even deeper into our realm of speculation, let's suppose such packets have a sort of dark matter structure analogous to and not unlike our own brains, mirroring their cognitive functions and activities. These dark shadow structures could have their own counterparts to our brain's white matter and nerve fibers, dendrites, axons, and neurons; all functioning on some form of dark energy that, just like dark matter, doesn't dissipate over time.

Science tells us that while dark matter and energy interact with all the visible matter and detectable energy that make up the universe, they do so in way that currently eludes our direct detection. Thus such dark entities are not only all around us, they can also fill every nook and cranny of our bodies and minds without us being aware of them. If that is true, the hypothetical "dark" soul packets just postulated could exist in parallel alongside the structures and functions of our biological brains without us having a clue that they're there.

Their undetectable activities could mirror and preserve all our cogitative brain functions. They would store our memories, experiences, thoughts, and the essence of our characters, and if this mirror image holds true for our brains it could likewise hold true for the rest of our bodies since such dark matter and energy would also infuse them and not just our brains. Then with our biological demises these soul packets would be liberated from whatever subtly binds them to our biological bodies and become free for God's transferal of them, or whatever parts of them prove useful to God, into the next plane of our existence, something we currently think of as heaven.

Is this pure science fiction? Who knows? However, it is intended to get us to consider the possibility that souls are real, and all the implications that would flow from that fact. As dark matter and energy-like substances they could exist indefinitely just as those two things have existed from the beginning of time without dissipation.

In a heavenly setting these soul packets would function as free-standing entities with, we believe, the capacity for sentient thought, self-awareness, and intelligent communication with their fellow souls, as well as with God. Of equal importance, they would be able to grow morally, spiritually, and intellectually.

But how will God productively employ those soul packets, us, in heaven? Remember, God is not our servant. We are God's servants. What will our souls do there that serves God's purpose and plan? Some will say, worship God! But it would seem unlikely that a God great enough and patient enough to create our entire universe would at the same time be so shallow as to need or enjoy a huge number of puny human souls endlessly shouting out their adoration of their deity. It also seems unlikely that God would have gone to all this trouble just so our souls could spend eternity lounging around heaven, basking in God's glory while God does all the work of running the universe and whatever comes after our universe's ultimate demise.

Also keep in mind that if our basic natures survive the transition into heaven, one of the innate characteristics God has engrained in each of us, as a result of our lives in this world, is our need to be occupied, to have something to do. The need for purpose is hardwired into the human physique and brain. More often than not, under God's tutelage, this need takes the form of a drive to be productive as opposed to simply engaging in activities that mindlessly pass the time away. Without being productive, of accomplishing something, of having a purpose, we become bored, uninterested in existence, and weary of the tedium of our lives on whatever plane.

So how will this need for productivity and purpose be met in heaven? How will God put us to work? What will God do in response to thousands of billions of souls wailing, "I'm bored" for tens of billions of years let alone eternity?

At least part of the answer to this question may possibly be found in the fact that God is a teacher and we are among God's students. Our review of history shows us God spending the entire span of *Homo sapiens'* existence teaching us the lessons God wishes us to learn. But, as of yet, we are clearly far from having achieved anything approaching a true mastery of God's precepts let alone living by them. During their earthly existence, the majority of our ancestors were even further removed from learning them than we are. So assuming all those ancestral souls will already be inhabiting heaven when we get there, it is a reasonable guess that some part of God will be engaged in fulfilling God's ongoing and chosen role of teaching them and us the lessons we haven't already mastered. The difference being that in heaven we will know that "God is" in all that those words can imply. Yet even with that knowledge, in such a setting, we will apparently still have to exercise our free will as we face, and seek to both absorb, those lessons and then act on them.

You may ask, "Why is God doing all this?" A possible guess is that God's efforts are pointed toward growing this entire universe of souls to the levels of near perfection it would seem God requires in order to be able to fulfill God's purpose and plan for all of our generations, past, present, and future. However, in terms of souls, we also need to ask if humanity is God's only focus? Or is that universe of souls made up of much more than just us?

When we talk about a universe of souls we cannot ignore the vast cosmos God has created that surrounds us. All those hundreds of billions of galaxies, each with their billions upon billions of individual stars and planetary systems. While we don't know for sure, it is only logical to surmise that God did not go to all the effort of creating such a nearly infinite physical structure and then drop only one tiny speck of sentient life onto just one of those planets, leaving everything else in that vast and endless realm sentiently barren and sterile.

Instead, we have to assume that there are other sentient races spread all across the universe's incredible vastness. Each of these alien species would be equally God's creation and equally loved by God. Therefore, we'd have to also assume that each individual member of each of those alien races possesses a God given soul, just as we do. And upon the biological death of those aliens, in whatever form God has created them, their soul packets will, likewise, transit to God's heavenly realm. So what is God's ultimate purpose and plan for their service to God? What does God have in mind for this unbelievably vast amalgamation of souls, human and otherwise? We can only wish we knew.

Chapter Seventeen

—··——··◇··——··—

TRUST AND PATIENCE

Since it would seem that one of God's major focuses has been on our education, perhaps we should address some of the lessons that God has spent so much time teaching us, but that so many of us often fail to recognize as being an important part of what God is concentrating on because these lessons may well serve as reflections of significant aspects of who and what God is. As such they can further advance our efforts to better understand God, despite all of the mystery of God.

Trust and patience are two of the foremost of those lessons that many of us tend to skip over when we think about God, but they are among the central themes God has spent enormous efforts instilling into every facet of human existence. Inasmuch as trust is a lynchpin for all our interpersonal relationships, societal structures, and interactions with God, let's look at it first. Without trust all of these fall apart. Almost everything that is positive in life requires trust as one of its key elements. Trust is therefore a critical ingredient of positive change.

This is especially true when it comes to our interactions with God. Since God has not seen fit to provide us absolute factual proof that a deity exists, our faiths in that existence has to be grounded in a trust that the promises to that effect are worth believing. God has worked very hard at teaching us to trust as a critical part of our lives, makeups, and character.

Jumping past the interpersonal implications of trust for the moment, and to which we will shortly return, think about the absolute requirement for trust in every aspect of our daily lives and the functioning of human civilization. As a quick example, when we board an airplane we trust that the plane has been well designed and well built, that it has been properly maintained and fueled, and that its pilots are competent to fly the aircraft and know where they are going. We likewise trust that the air traffic control system and the human air traffic controllers behind that system will keep all the thousands of planes that are in the sky at any given moment safely separated so that they won't collide with one another.

Without that trust the passengers on Malaysian Flight 370 would never have gotten on that ill-fated flight in March 2014 when something went terribly wrong. Instead of flying to its intended destination in China, it went in the opposite direction and disappeared somewhere in the Indian Ocean off Australia. Something or someone caused that flight to violate the trust of its passengers at the cost of their lives.

Despite that violation, millions of other passengers have continued to board other planes all over the world month after month since then. They did so because they still trusted the overall aviation industry. Why is that? It is because we have had the concepts of trust ingrained in us by both our experiences and by God.

Similarly, we have learned to trust that our commercial systems will ensure that a vast network of agricultural and food production efforts by hundreds of millions of our fellow humans all over the world will come together, day after day, to ensure that "our daily bread" is always available to the vast majority of us. Those who are Christians are comfortable praying to God, as part of the Lord's Prayer, to "give us this day [that] daily bread," because God has taught Christians to do so and most of them are confident that this prayer will be answered.

So it comes as something of a jolt to many of us to be reminded that there are large numbers of people around us and around the rest of the world who don't have enough to eat each day to sustain their lives. Nonetheless, based on the trust God has taught us, when that reality hits home we don't all turn into food hoarders whose fear and lack of trust would collectively bring the entire food distribution system crashing down.

156

God's efforts to instill this trust in humanity go back to at least our earliest recorded history. Many of the stories God has inspired in the Bible revolve around what it takes to teach humanity to trust God and each other. These are the tales that have infused and informed our societies' very structures. To see this we need look no further that the Biblical saga of the Israelite flight out of Egypt and the forty years we are told they wandered in the desert. Despite wonders depicted in the Book of Exodus as including God's parting of the Red Sea, using pillars of smoke by day and fire by night as beacons to guide the Israelites in their journey, producing water out of solid rock, and feeding them with manna from heaven, the rabble that had fled Egypt under Moses' direction couldn't bring themselves to fully trust in God. Instead, they bickered with each other and Moses, questioning every step of the path along which God is portrayed as leading them.

God's response was to patiently teach them to trust. We are told that God first gave the Israelites the Ten Commandments and then the rest of the Laws attributed to Moses. Thereafter, God spent millennia pounding trust of those laws into God's chosen people, as well as the need to cooperatively follow them as a society. In that process the descendants of the twelve tribes of Israel forged their trust in God and each other into something that has served them throughout their history, right down to today.

Through the story and example of their migration, this lesson has been preached to Jews, Christians, and Muslims ever since. It helps form the backdrop that underpins the concept of trust found in every society that flows from these religions and by extension the rest of the world. It also helps to demonstrate that people must learn to trust in and live by valid laws, as well as trusting in God.

Trust is likewise critical to all of our successful interpersonal relationships, be they in marriage, families, the workplace and business, or in simple friendship. And it is especially important in our interpersonal relationship with God.

Starting on the human level, each of us has to trust the individual we chose to bond with in anticipation of spending the rest of our intimate lives in their company. We place our emotional wellbeing in that person's hands, so to speak. No matter how well we think we know that individual, it takes real trust to make such a commitment. God understands this. So a significant part of God's teaching efforts have focused on not only love, but also on building up the human emotional capacity to trust.

We learn these lessons of trust from our infancies. We have to trust that we can depend on our parents and extended families, over a protracted period of years, for our emotional wellbeing as well as for the most fundamental of our needs – the shelter, food, clothing, and education we require to grow into mature, productive, and caring adults. This effort also requires our trusting that the larger society of other families and social groupings surrounding our nuclear family will likewise assist and protect each other and us in these efforts. Without such trust those families and societal groupings would fragment and fall apart.

Since God prizes our growth, as individuals and as societies, God has consistently taught us the lessons of trust required to form the heart of all of these relationships. It is what is critical to the advancement and expansion of God's plans for all of us. It is a constant in all that God does with humanity.

This requirement of trust extends outward past the family into all of our workplace and business relationships. It cements the commerce required for all of our economies, starting at the local level, and then broadening into ever larger circles and an interlocking whole that is necessary for humanity to function successfully. Thus, trust has to exist throughout all of our societies at every level until all of humanity becomes one vast network of interdependent economic structures functioning on at least minimal levels of trust. Without a basic level of trust this entire latticework would come crashing down leading to the collapse of human civilization.

Throughout all this, God has patiently worked at instilling notions of trust into all of humanity so such a collapse won't happen. At the same time God was teaching us how to trust our immediate families and societies, God also initiated the effort to lead us toward trusting in God's own self. When we peruse the Bible, we find the instruction to "trust in God," to "trust in the Lord" and their variants all through its pages. They are found from beginning to end in both the Old and New Testaments. In book after book, the majority of the Bible's human authors, acting under God's silent and subliminal urgings, incorporate this admonition hundreds of times. We are enjoined to do so more than fifty times in the Psalms alone.

Nonetheless, when thinking in terms of trusting God many of us conjoin the phrases trust in God and faith in God. People who do so seem to be thinking of them as being the same thing, but they are not. They are two different acts. We have faith that God exists. Based on that faith we place our trust, that is to say our confidence in God, trusting that God will act in ways that are in humanity's overall best interest. This is the concept God has expended so much time and effort on teaching us.

It is a relatively logical step, although not necessarily an easy one, to go from the greatest commandment, loving and trusting God, to that of trusting each other just as we are commanded to love each other with the biblical admonition to "love our neighbor as ourselves," which we are told is the second greatest commandment to be found in the Bible. But we should recognize that it is God who is leading us along this path because, as we've just seen, this is what God wants us to do as a critical component of our growth and advancement.

Many of us seem to be coming to understand this. Thus a majority of the people who have begun to more or less believe in God, in whatever form they conceive of the deity, also believe that God is trustworthy and has our best interests at heart. They have in turn transferred this learned trust to the rest of the human institutions and relationships that enfold their lives unless and until those relationships prove them wrong.

In fact, the concept of trust has become so ingrained in our physiques that when, say, an imam, priest, rabbi, or minister of the cloth breaks the vows of love and service that bind them to their flocks, people are dismayed and shocked by their actions. You need look no further than the epidemic of pedophile priests whose violation of their vows have in recent years stunned the worldwide Catholic Church, but despite these widespread abominations by an admittedly small minority of the priesthood, the majority of the Catholic faithful still adhere to a basic trust in their Church and God, even as they demand reforms and restitution to the victims of these criminal violations of what was a trusting relationship.

The same phenomena holds true all across the human spectrum. We remain committed to trust our democratic institutions and forms of governance, even when we are confronted with a parade of elected or appointed officials at all levels of government in every nation who breach our trust through either graft, corruption, or a preoccupation with their own narrowly focused self-interests, which they pursue to the detriment of the greater good. The concepts of trust have become so ingrained in us that even when an endless line of individuals break a trusting relationship, be it in personal relations, in business, or in social or political contexts, we still tend to look for others we can trust in the same arenas.

So what does this need tell us about God? First, that God understands our needs. And by extension, God understands that if God's own self was malevolent or even just capricious, we humans would end up not trusting God or each other. That is clearly not what God is aiming for in terms of our relationship with God and with the rest of humanity. It would not advance God's apparent plans for us.

Therefore, God is teaching us to mold our relationships with each other on the relationship we should have with God based on loving trust. This model is founded on trust precisely because trust is a basic and necessary ingredient for almost all positive human interactions, advancement, and change. God is intent on that sort of positive trusting advancement for all of us in all of our generations. This especially includes our need for ongoing trust in whatever awaits us after our earthly lives. In fact, it may well be a key part of God's efforts to advance our growth in God's heavenly realm.

It also tells us that God wants us to be able to trust each other and to come to understand that ultimately we can, out of our own free will, trust God. It likewise points to trustworthiness as being one of the basic traits of God's own makeup and character, and this is a trait God is leading us to mirror.

Which brings us to the next lesson: patience, one in which God seems to be having less success with humanity. To put this into perspective, it may be useful to compare God's patience with modern human patience.

We have already described the 4.5 billion years God has spent patiently evolving our solar system and Earth with all its life. This most recently includes the two hundred thousand or so years God has been working on humanity's advancement. Thus, by definition, we must acknowledge God's undoubted patience. Obviously, God is in no hurry. But then God is an infinite being working in eternity, so patience isn't necessarily a needed virtue.

We humans by contrast, are short-lived creatures who are constantly in a hurry. We are astonished when we encounter what we think of as unbelievable human patience. For a great example of this, we need look no further than the brilliant work of Catalan architect Antoni Gaudi (1852 – 1926) to see one of the most incredible manifestations of human patience available to us.

Gaudi lived and worked his entire life in Barcelona, Spain. Even though he has now been dead more than ninety years, his designs and buildings are still considered to be some of the best and most advanced architectural work ever done. Tourists and young architects take tours of Barcelona and stand in lines just to see his buildings and a park he designed. His masterpiece, however, is the Basilica of La Sagrada Familia. The project for this church in Barcelona had only been underway for a year when, in 1883 at age thirty, Gaudi was selected to be its architect. He would spend the remaining forty-three years of his life working on it as one of his principal focuses. In fact, he spent the last ten years of that life working on nothing else, even to the extent of living under austere circumstances in his design studio in order to avoid other distractions.

However, even while doing so, Gaudi knew he would never live to see this work completed. In fact the construction of the Basilica's exterior facades, architectural fantasies in themselves, are still not finally finished, and are not expected to be done until 2025, one-hundred and forty-three years after they were started. Its interior spaces, though, were completed in 2010, and the church is now open to the public for worship and inspection. The results are breathtakingly beautiful. The Basilica's incredibly open vaulted ceilings seem to reach to the heavens. Light floods into the sanctuary from all sides through enormous stained-glass windows, each of an abstract design with individual color schemes. The totality is an ethereal, uplifting masterpiece that stands with the world's greatest edifices.

If you were born in say 1939 this building would have already been under construction for fifty-seven years and chances are you will be lucky to see it finally finished in your lifetime. Generations of craftsmen have spent their lives in this effort; thus, it is the product of one man's incredible patience and genius, as well as the combined efforts of a small army of people dedicated to completing it after Gaudi's death. In this sense the Sagrada Familia is a product and example of the patient multigenerational human effort needed to produce the positive changes that advance our civilizations and societies.

Their accomplishment can only be considered awe inspiring. The concept of taking nearly one hundred fifty years to finish just one building boggles our minds. Yet even as people worship God in Gaudi's Basilica, some never stop to hold Gaudi's timeline up against the time frames God, who has been working on our planet and us for so long, lives by.

Part of the reason we miss this connection is that so many in our societies adhere to constant refrains of "Is it done yet? When will we get there? Hurry up, I'm bored!" We haven't learned the patience needed to appreciate the Gaudies of our world, let alone the patience of God. But it is exactly those forms of patience God has been trying to teach us.

In response, there are any number of people who haven't been listening to God, let alone learning God's lessons in patience; instead, they are controlled by impatience and want to project that impatience onto God. They attempt to dictate to God as to when and how fast God should act, and when God doesn't heed them, they are tempted to show God how things should be done. For example, driven by fervor and impatience, whole groups declare that those who worship God in ways that differ from the first group's exact form of worship must be evil and deserve death. They aren't willing to patiently wait for God to make that judgment when those other people finally appear before God at the end of their relatively short earthly lives. The foremost example of this impatience is the bloody split in Islam between the Sunnis and Shias; the most extreme of whom rush to hurry death along for the other sect's adherents, even as they both profess to love God and God's teachings.

Watching this conduct, the rest of the world wonders at God's indulgence of this sort of mayhem. However, God's patience in this regard isn't to be equated with indulgence in the sense of giving in to willful children when they misbehave. What God is first doing is silently urging us to deal with these problems as part of our growth. If we fail to do so, God will ultimately address the problem. We probably won't like how God does it, especially when people can't muster even the tiniest fraction of the patience God has demonstrated.

163

Chapter Eighteen

PARTNERSHIPS IN SHADES OF DARKNESS AND LIGHT

Another of God's lessons deals with our partnering with God. When it comes to influencing the human condition and human conduct, God has demonstrated a preference for incorporating mutual partnerships into mankind's interactions with God. This is especially true when it comes to ameliorating the darkness of the human condition.

In these partnerships God is, more often than not, the guiding and sustaining force, pointing us toward the paths we should follow and helping us understand the actions that need to be taken. They are not, however, partnerships in which God does the actual, physical work. Nor are they partnerships among equals, or even near equals; instead, they are a connection of grace between the infinite that is God and the minute that is humanity in its individual parts. It is only possible for this, oh-so-unequal, relationship to come into existence because of the love God has for us, a love which we can only return in miniature measure from our individual souls and personal love of God.

In these endeavors a metaphor proposed by Rabbi Mitch Chefitz can serve as a reflection of our relationship with God. Mitch talks about the awe and elation he felt one morning as he sat on a finite point of land in the Florida Keys watching a perfect and gloriously spectacular sunrise unfold before him. He felt himself bathed in its rays as the Sun's golden red orb lifted out of the ocean's predawn gloom into the sky in front of his eyes.

Then it struck him that the rising of the Sun was an illusion created by the Earth's rotation. The Sun is where it always is, in the center of our solar system. It spreads its life-giving warmth uniformly across our rotating planet all day long each and every day. But we only see it when our little spot on this Earth comes around to face the Sun at a time when storm clouds, fog, or atmospheric pollution don't obscure its glory from our view. However, even in those moments when we can't see it, the Sun's rays still penetrate whatever floats between it and us to warm and nourish each of us if we would only recognize its presence.

Just like the Sun, God is always there at the center of our lives washing the divine essence of God's being over us. God's own self is lovingly available in partnership to work with us from God's limitless presence to our infinitely smaller and finite point of existence.

Using our free will, we have to push through whatever obscures or blocks God from our view and either elect to enter into this unequal, but loving, partnership or reject it. In oversimplified terms, and using the words of Benjamin Franklin, "God helps those who help themselves." But helping ourselves, more often than not, first entails our opening ourselves to God's reaching essence, subliminal urgings, messages, and instruction.

As we do so, we have to seek for and listen to God's guidance and directions. This especially includes God's admonitions when we are wrong or when God's answers are either inconvenient or entail sacrifice on our parts. Then we have to act on what we are hearing as it wells up within us, no matter how much we might not like God's response.

This opening to God operates on many levels, but they all begin with each individual person. The most basic of these levels is found in our own bodies and minds. All of us have a vast array of systems within our bodies and brains that regulate the various functions that are critical to our health and wellbeing. Most of these are involuntary and autonomous: our immune system being just one example. When one or more of these systems malfunctions or is attacked we automatically think of seeking the aid of human medical science, which is fine. But, when we do so, if we will at the same time turn to God and open ourselves to God's healing graces, surprisingly positive results often flow from that partnership. Of course, this does not mean we can ask to live forever in our corporal bodies. But it can produce enhanced experiences and outcomes in the life we are given.

This same individual partnership also exists in the realm of our emotional health and sense of personal worth. Day in and day out, we each have to live first with ourselves and deal with our own limitations, insecurities, and neuroses. We have to come to grips with our individual fears and personal challenges. We don't, however, have to do so alone. When we allow our loving God, who knows us far better than we do ourselves, to participate in our efforts to handle each of these issues, we will once again frequently find positive outcomes that will both surprise us as well as lead to far greater internal peace. We just have to give the ever available partnerships with God a chance to work instead of throwing up mental and emotional walls that block God out.

But even as we cope with our own individual personal issues, we also have to deal with the questions that arise from our families and communities. This circle then widens out to take in the greater social, political, religious, and commercial entities we interact with and live in. Ultimately, this includes the entire world with all its nations and various human groupings since, in one way or another, their actions end up affecting us just as we affect them.

This all requires an almost endless number of effectively functioning partnerships on the human level. And those partnerships become truly productive when we individually and collectively open them to God's participation as well. As we partner with the divine, God will also be urging our fellow humans, who are joining in this opening to the Father, to accompany us in taking on the tasks of finding the solutions required in order to deal with the common dilemmas we all face across all those partnerships. In doing so, God will likewise subliminally suggest to all of us a range of possible answers to our questions.

By partnering in this way with God and our fellow humans, we are more likely to discover effective resolutions of these issues, threats, and dangers than if we bar God's input into the equation.

What we can't do, however, is blithely consign to God the dilemmas and threats humanity must overcome. Doing so is not a way of absolving ourselves from the obligation of taking them on. Nor can we expect others to assume the responsibilities of partnering with God in our stead, leaving us free to enjoy the fruits of God's and those other people's labors. The issues facing humankind are now so pervasive and all-encompassing that it will take the combined efforts of both humanity and God to master them.

In the same way, in times of desperation it does no good to simply shout at God, pleading for divine intervention to protect us. That's a one-way street that defeats the very purpose of partnering with God. Instead, we have to master the art of opening ourselves and being still in God presence, as we listen for God's silent voice and instructions. With that opening we invite God's participation in marshaling our individual and combined human efforts along avenues that point us toward the solutions we are groping for.

The humanity-wide menaces we must all collectively face have now grown so immense and intractable that they are threatening the very existence of our combined civilizations. Our efforts to understand God's plan thus far suggest that at this point in our history of evolutionary growth God will no longer unilaterally intervene to save us. Instead we must join in being active participants in the required solutions. We have to be the moving agents who execute whatever answers God helps us come up with. We are the real-world actors who effectuate the changes required to overcome these menaces. This effort is an essential part of the growth God is pushing us toward in our evolutionary climb out of the animal realms and onto the road to the future God has in mind for us as sentient beings.

However, the question becomes will we address these dangers early on, or only at the last minute when they are about to overwhelm us? In this the twenty-first century the primary source of these threats is each and every one of us living today. We are the authors of the darkest of the calamities boiling up on humanity's horizon.

That's because all of these threats arise out of our collective, heedlessly shortsighted insistence on satisfying our individual, evolutionarily inherited animal appetites, needs, and certainties. We are at such levels of excess that they have become massive and destructive addictions. Far, far too many of us are currently pursuing this quest for selfish self-gratification at the expense of all the rest of humanity and all its future generations, as well as denying our potential partnerships with God.

The first of these addictions is population growth. Today, in aggregate we are a mass of more than seven billion people, a figure that grows daily. Seven billion, a seven followed by nine zeros, is a number that is hard to get our minds around. The concept of that many people is difficult to grasp emotionally. One *National Geographic Magazine* article on our world's population attempted to put this number into prospective by suggesting that if any one of us tried to count from one to seven billion with each beat of the count taking just one second it would require more than 200 years of 24/7 effort to complete the task.[45]

And seven billion plus is only the population count we passed through in 2013 as part of our evolutionarily driven urge for procreation – an animalistic urge that drives us toward higher and higher numbers of people all seeking the same things. The current best guesses of world demographers are that, if we're lucky, sometime between 2045 and 2055 we will, hopefully, top out at nine plus billion people sharing this lifeboat of a planet. At that point population scientists predict humanity's fertility and reproductive rates will decline to a point where, on a worldwide average, we will reach what is known as the replacement plateau. There we will only be adding enough new souls to replace those who are departing this life.[46] Of course, that is only thirty or so years into the future.

If we don't achieve this plateau until say 2065, that forty-six years from now, the globe's population will have swelled to a staggering fourteen billion people, twice what it is today.[47] If we never reach such a plateau our uncontrolled reproduction will lead to around twenty eight billion people, another doubling, by 2100, less than eighty-five years into our future. All this will either threaten the collapse of human society or the quality of human life will be so degraded that God's patiently taught lessons will be lost. Our Earth can simply not sustain such numbers.

The less economically developed portions of our world will account for 95% of this population increase. Many of the planet's more industrially advanced countries including China, France, Italy, Japan, Russia, and others are already at or near reproductive levels that have dropped below their replacement ratios.[48]

Nonetheless, it is the interlocking impacts of our worldwide numbers that will generate unsustainable pressures on Earth's environments, climate, and resources, as well as on each of us individually, that will be critical. Our worldwide population growth will produce cycles of negative change that will challenge us as we've never been challenged before. How we deal with these issues of change will dictate humanity's future and how God deals with us in either the light of the partnerships we should be establishing with God, or the darkness that will exist if we do not.

The potential consequences of all this dictate that we must quickly focus on the combined impact that seven, nine, fourteen, or twenty-eight billion people will have on the world's climate, its food stocks, and the available fresh water supplies, as well as on its fossil fuel consumption and reserves and other nonrenewable resources. What will happen when each of these tens of billions of souls adds their individual consumption demands on those finite resources all the while contributing to the greenhouse gas load Earth's atmosphere is forced to bear by humanity's cumulative usage patterns? What kinds of negative change will result? Because it is out of these numbers that the rest of the threats we will discuss emerge.

Our Earth's rapidly expanding population, coupled with the significant short-term economic gains being achieved by large segments of the world's various nations, is currently fueling worldwide climate change in the form of global warming. Our individual-by-individual consumption of energy resources be they in the form of gas-powered engines, coal-fired electric power plants, wood-burning fires, or a host of other human-generated sources, climbs ever higher. And they all add to the buildup of greenhouse gases as we pour more and more resulting carbon into the atmosphere. The consequence of that increase is the seemingly inevitable decade-by-decade march of higher and higher average world temperatures. And their rise will force change on all of humanity, on each of us.

The consequential climate change that will result will produce a situation in which we will face both too much water and not enough. The too much side of the water coin will be found in the rise of worldwide sea levels over the span of the twenty-first century. The world's rising atmospheric temperatures are thawing vast amounts of water previously trapped in snow caps and mountain glaciers, as well as the ice dome covering Greenland, and Antarctica's ice shelves.

How much sea levels will rise is a matter of widely varying conjecture depending on which set of facts come to pass. And those facts will be heavily influenced by what we all do individually and collectively; as well as whether we are partners with God in finding solutions or ignoring God's proffered assistance in marshalling the humanity-wide cooperation needed to alter our rates of procreation and patterns of resource consumption. If, over the next hundred years, half of Greenland's ice dome and half of Antarctica's ice shelves were to either melt or break up and slip into the world's oceans, those seas would rise between eighteen and twenty feet.[49] The most recent studies of the West Antarctic ice shelf indicate that this floating mega mass of ice is melting from the bottom up as a result of our warming atmospheric conditions that are heating up the ocean currents flowing under that shelf's ice fields.[50] At the other polar extreme, Greenland is losing its ice cover at an alarming rate. Its glaciers are visibly receding year after year with waterfalls of melting runoff pouring off them to end up in the ocean.

Absent drastic action on our collective parts, our children and grandchildren, many of whom have a statistical probability of living into the twenty-second century, less than eighty years from now, will have to face the consequences of our selfish inaction. A just-completed analysis determined that more than 630 million people currently live in vulnerable coastal areas whose land elevations are less than 30 feet (9.1 meters) above sea level. This includes two thirds of the world's cities with populations of five million or more.

While this is a worst case scenario and predicting climate-change induced sea level rise is an inexact science, at the very least sea levels will almost certainly increase by a minimum of two to four feet (0.6 to 1.3 meters) during the next fifty to one hundred years.[51] The unprecedented shrinkage in Greenland's ice dome and the collapse of major chunks of the West Antarctic ice shelf, in conjunction, have initiated exactly this sort of measurable oceanic rise.[52]

Even these less catastrophic increases in ocean heights will have major negative impacts on the world's coasts. Humanity's future generations will not only deal with the results of greater flooding produced by rising sea levels, but they will also have to confront massively amplified destruction events in newly low-lying areas when those areas are pounded by hurricanes and winter storm activity not unlike 2012's Hurricane Sandy that embedded itself in a nor'easter, devastating the mid-Atlantic coastal areas of New Jersey, New York, Connecticut, Road Island, and Massachusetts as well as the Canadian Maritime Provinces.

And lest we think of this disaster as a one-time event we need look only to 2016's Hurricane Matthew, 2017's Hurricanes Irma and Maria, and 2018's Hurricane Michael. All of these were incredibly devastating category 5 storms.

Most recently 2019's category 5 hurricane Dorian with its 185 mile per hour sustained winds literally wiped out the northern Bahama's

Abaco and Grand Bahama Islands along with their towns of Marsh Harbor and Freeport. In its aftermath tens of thousands of their island-ers have become environmental refugees and the loss of life was enor-mous. One can only shudder to think about what the destruction would have been had that monster storm continued on its westward track and hit southern Florida instead of veering northward to rake the east coasts of the U.S. and Canada.

Too much water, however, is just part of the H2O picture. The other part is too little water for the life-sustaining needs of nine to four-teen or more billion people. This could be an even bigger problem.

Without question the world's glaciers are shrinking at an alarming rate under the relentless onslaught of our rising atmospheric tempera-tures predominately generated by humanity's voracious demand for fossil and biological fuels to produce energy, along with that energy's unwanted offspring, greenhouse gases. The poster child for the resul-tant glacial thaw can be found in the vast expanse of ice fields and snow caps spread across central Asia's Himalayan Mountains and the Tibetan Plateau. This collection of glacier fields and snow packs is sometimes referred to as the Earth's third pole. Draining off this mass of ice and snow are seven of the globe's greatest rivers. Starting from west to east they are the Indus, Ganges, Brahmaputra, Salween, Me-kong, Yangtze, and Yellow Rivers. Nearly forty percent of the world's entire population depends in significant part on these seven rivers for their freshwater needs.[53]

Initially the increased rates of glacial melt water runoff from the Tibetan Plateau and the Himalayas will produce an overabundance of water, inducing floods and major erosion problems. Eventually, how-ever, at some indeterminate point twenty to fifty years from now de-pending on a host of different conditions, there will be a steep falloff in the volume of water flowing down those seven giant waterways as their frigid sources shrivel and recede under the attack of global warming.

The implications of these diminished water flows, while the pop-ulations that depend on them continue to swell, are frightening. There will be acute shortages of both fresh water and hydroelectrically generated power that previously fed off these rivers. Food production could well plummet. Mass movements of ecological refugees are a real possibility, especially from the countryside into already overburdened urban areas. There may well be military conflicts and wars between the Asian nations, who currently have to share these water resources, as to who gets and controls the water that remains in these water systems.

At the same time the rest of the world will be eyeing their own water woes. Israel, Syria, Jordan, and the West Bank Palestinians will all want the waters of the Jordan Valley. Nations such as Turkey, Syria, Iraqi, and Iran could be at each other's throats over who controls the Tigris and Euphrates Rivers plus their tributaries. Vast portions of the Earth's subterranean aquifers, which took millions of years to fill, will have been drained by human consumption and irrigation far faster than they can be replenished by nature. This fact alone could have disastrous consequences for world agriculture and the human food supply system needed to feed upwards of nine billion people.

Even in countries that don't have to share their river systems with their neighbors, local regions will fight over water. We can already see this today in terms of Northern and Southern California, Nevada, and Arizona jockeying over how much each gets out of the Colorado River and the mountain runoff from the Sierras. Mexico used to get some of this water, but its current ration is down to a trickle. In the southeastern US, Georgia, Florida, and Alabama have an ongoing battle over how much water can be taken out of the Apalachicola River for the needs of the city of Atlanta at the expense of the two adjoining states.

On a less apoplectic, but nonetheless devastating, level will be the shifts in the Earth's weather patterns. Our increasing average global temperatures will set new record highs on a regular basis. This will inevitably lead to prolonged heat waves that will result in the deaths of tens of thousands. In 2003 just such a heat wave killed 35,000 people in Europe.[54] And as our oceans get warmer each decade from this temperature buildup, the stored heat in their waters will fuel more and fiercer hurricanes, typhoons, Indian Ocean cyclones, and monsoon-like rain systems. There will generally be more severe weather outbreaks including more deadly tornadoes and massive increases in local flooding.[55]

These shifting weather patterns will likewise cause droughts and desertification in areas not previously prone to such problems. Not unlike the prolonged drought that struck the central US from the Rockies to the Appalachian Mountains in 2012 and Southern California in 2013-14, these conditions will decimate wheat, soybean, and corn harvests and destroy pasture lands requiring the selloff and slaughter of herds of livestock. When such conditions continue over not just years, but decades, they can become killers of entire communities and societies. The resultant famines will mirror what has already occurred in the Sudans, Somalia, Ethiopia, and other parts of Africa including the Darfur region. At the very least they will massively drive up food prices on a worldwide basis and impoverish large populations.

The next group of threats – religious intolerance, racism, bigotry, and self-interest – are nasty words, but they don't initially suggest the immediate impacts on humanity of dangers such as endemic famine, massive flooding, or killer thirst. Their end results, however, can be just as devastating. And all four will rear their ugly heads on a worldwide basis if our inactions in the coming years lead to global sea-level rise-induced flooding of coastal zones, drought in inland areas, drying up river systems, and expanding desertification.

By themselves and in combination these disasters, when spread across the globe, will wreak havoc on major portions of humanity. They will produce waves of environmental and political refugees. Costal populations will be forced inland toward higher ground. Deprived of enough fresh water, agricultural production will fall off in many areas, leading to food scarcity migrations as people seek lands that still have enough moisture to grow crops. Rural populations, which can no longer sustain themselves locally, will flock to urban centers in search of work, food, and fresh water. Trading patterns will be disrupted as nations begin to hoard their own food stocks or charge exorbitant prices for such food as they are still willing to sell.

In the face of these forces, people will once again start looking for justifications as to why they should stop considering newly displaced populations, (many of which will be of different skin color, race, or religion) as fellow human beings who are deserving of help. Instead, absent God's partnership-induced intervention and teaching, there will be a growing tendency to see such people as threats to anyone who still has what those refugees want and need, be it water, food, arable land, jobs, or just survival and safety. A host of latent prejudices and animosities will quickly resurface, and those already in the open will be exacerbated.

Pushed by these stresses, the religious fault lines between groups like the Sunnis and Shiites will flair into even darker and more wide-spread conflict, as will those pitting Islam against Christianity, and Judaism. Old Hindu, Islamic, and Buddhist animosities will again boil to the fore. The conflict between the cultures of modern western civilization and retrograde fundamentalist jihadists will become exponentially more deadly than they already are. Rwandan-type tribal warfare will become more the norm as opposed to being the exception. And the have versus have-not conflicts may become overwhelming.

The 2014- 2016 mass movement of streams of desperate people frantically seeking to force their way into Europe may have already given us what will seem, in hindsight, to be a harbinger of what looms before us. This flood of humanity fleeing war, drought, sectarian conflict, and poverty in countries as diverse as Syria, Iraqi, Afghanistan, Pakistan, and the Sudans has so far numbered nearly three million souls, with no end in sight.

These people have spent almost everything they had on perilous water crossings or overland smuggling routes from Turkey, Lebanon, Egypt, and Libya in which many have perished. The TV footage of this seemingly endless horde of people jammed together in dilapidated boats and rubber rafts, or being pulled dead or alive from the sea after sinkings and wrecks is mind-numbing until you see the lifeless body of one little toddler cradled in the arms of a would-be rescuer.

The images of never-ending lines of forlorn young and old men, women, and children trudging across the Balkans only to come up against newly erected fences barring the borders of country after country who are equally desperate to keep from being overwhelmed by this flood of humanity is heart wrenching. The stress on countries like Turkey, Lebanon, France, Greece, Italy, the tiny Balkan states, and the rest of Europe as they seek to cope with this tidal wave of humanity, not to mention the possibilities of them serving as cover for terrorist and their attacks, is almost incalculable.

Today we are seeing the same horrific story playing out on our border with Mexico. There hundreds of thousands of migrants fleeing similar problems in their Central American countries are seeking refuge in the U.S. Now it's just closer to us but with the same terrible scenes including the picture of a father and his little daughter drowned attempting to illegally cross the Rio Grande.

These unwanted migrations have opened all the fault lines of racial and religious and class intolerance. It also has the potential to create mass political instability, along with cultural change in the countries being inundated by these waves of refugees. Its effect will be a true test of how God's lessons will be implemented. As the rest of the world watches all this, we have to recognize that it represents not just a European or American problem; instead, it is a world problem that all of us are going to need God's help in dealing with. When you magnify this ten, twenty or a hundred fold you can begin to imagine what the future may hold for the world if all of us don't jointly address the challenges facing us today.

These are just the top of a laundry list of the problems that humanity is going to have to work through during the lifetimes of everyone living today and that we are bequeathing to our children and grandchildren. The question is, however, how are we going to address them?

Will we do so early on with God's help and partnership, or will we wait until the last minute to acknowledge their true magnitudes. And when we do, will we then cry for divine help that may not be forthcoming in the ways we would hope because of our own inaction.

The answers to these questions are up to us. For God has given ample evidence that God is not going to step in and do all the heavy lifting while we continue to wallow in our own self-indulgent and narrow-minded attitudes toward the rest of humanity and God. Retreating into denial of the hard, but obvious, facts and issues because they demand sacrifice and difficult choices is the road to blocking out God's help and partnership. It leaves us with only our own devises that have so far been inadequate to overcoming these problems. If, bowing to the darker animal instincts God has been so patiently growing us out of, we give in to blind self-interest and selfish responses to all these threats at the expense of others, we cannot expect God to salvage the situation for us at the last moment.

If on the other hand, we individually and collectively open ourselves to partnerships with God, and listen to God's silent instructions, we will find many of the answers that will ameliorate most if not all of these threats. But sooner or later, in darkness or light, God will ultimately force us to confront all of these failings in our human character and weed them out. We can do it the easier way in partnership with God or the rock hard way without God, but confront them we will. Let's just hope God continues to think we are worth salvaging at the end of this process.

And just as Pope Francis so recently reminded us in his encyclical *Laudate Si,* we can never lose sight of the fact that God has given us the stewardship of both our planet and each other. It is up us to nurture and protect both.

Chapter Nineteen

GOD IS THE ULTIMATE TEACHER

Throughout our discussions we have noted God's ceaseless efforts down through the ages have been to teach humanity a series of possibilities, lessons, concepts, precepts, and attitudes. These include caring, gratitude, sharing, forgiveness, and empathy among others. They also count among their number trust, fidelity, morality, honesty, restraint, obedience, self-sacrifice, and the common good. Capping all these are the overriding concepts of reward and punishment for good conduct or bad, faith, love, and the possibility of eternal salvation. In short, all that is central to growing us out of our historic past as selfish, self-centered animals, into rational functioning individuals and societies capable of partnering with an eternal God.

As we consider all this it becomes obvious that one of the principal aspects of God, which we can both perceive and interact with, is that of God as a loving teacher. For God is a teacher, the ultimate teacher in fact. The corollary to this is that we must therefore be God's students. And just like students in any school, we must first enter God's earthly classroom and then eventually leave. But since this is the school of life, we – its students – enter at birth and only leave, generation after generation, with death. But, hopefully, our departure from this life is only a necessary step toward proceeding onward to graduate school in heaven.

As our teacher, God has had to gage and test each generation of students God has interacted with. This required God to evaluate each cohort's capacity to absorb the lessons being imparted to them and now to us; likewise, God has had to present lessons in formats each generation could understand. It would do neither God nor God's pupils any good for God to have attempted to teach humanity complex concepts and factual realities when the class didn't have the underlying intellectual and knowledge basis required to comprehend what was being taught. This would be like using Boolean algebra, differential calculus, and quantum mechanics to communicate with pupils who hadn't yet mastered basic arithmetic.

Keeping this in mind, and looking as far back in time as we have any sort of historical records, we can see God's students struggling to understand the lessons being patiently imparted to them within the parameters of their abilities to comprehend God's subliminal teachings and messages. Some of those students in turn became the authors of the stories and texts God would later use to teach subsequent generations. The Bible and Koran are two of those texts, and some of their stories such as Adam and Eve, Noah, and a number of others were all couched in terms that, over the last four thousand plus years, God's student readers could intellectually understand, remember, and learn from even if they weren't literally true. This was the function of those stories as opposed to being factual history.

So when God the teacher was inspiring God's students to write such stories and then teach others from them it was only in the last several hundred years or so that God's pupils began to achieve the intellectual and cultural bases required to just begin to comprehend the time frames involved in the evolution of life here on Earth, not to mention the vastly greater time needed for our universe to evolve. Understanding the physics, chemistry, biology, and math associated with all this would have been far beyond anything most of our ancestors could have even imagined, let alone understood. Nonetheless, God had to teach them something that would serve God's purposes as humanity edged out of its animalistic past and into its humanistic future. Therefore we now have the fables and parables we and the generations before us have grown up with – ones that have served so many of us so well.

179

Despite these impediments, over the millennia these same students have advanced, little by little, in cumulative knowledge and sophistication to the point that today they can recognize the difference between factual history and allegorical teaching tools that served to implant the commands, concepts, and ideas God the teacher would have us absorb and live by. But in doing so God doesn't want us to throw those God-inspired stores into the dust bin.

As a teacher, God still uses them to educate and inform the hearts, minds, and attitudes of God's human pupils. But, ultimately, it is we the students who must act on what we have learned. Just as in every school that any of us has ever attended, it is the student who has to leave the cloistered halls of learning and go out into the real world to utilize and apply the internalized lessons we have come to understand as the core values associated with living our lives. That's why God has taught them to us.

Like rebellious teenagers, we can sit in God's classrooms and tune God out, because God is interfering with our pleasures, asking us to do difficult things or just boring us, or we can wake up and listen, learn, and then apply God's lessons in our lives and dealings with the rest of the world.

Critical to all this is the recognition that the rest of that world and its experiences are integral parts of God's teaching methodology. In fact, that world is the living experiential laboratory that God the teacher uses to drive home God's lessons. For God understands that it is one thing to intellectually comprehend in an abstract sense the darker sides of life as well as the sublime – things like hunger, lust, greed, fear, pain, anger, hatred, bigotry, racism, and all the rest of the ills and temptations the world suffers from. It is an altogether different one to have to acquire a real-world knowledge and intimate familiarity with them as a result of having encountered them in the flesh.

This lesson was brought shockingly home to the world on February 3, 2015, by ISIS's release of gruesomely graphic videos of its execution of Jordanian Air Force F16 combat pilot, Lieutenant Moaz

Kasasbeh. In stunned horror, viewers saw a living twenty-six-year-old man burned to death locked in a cage. His Jihadist captors squatted around that cage watching approvingly as this hideously barbaric act unfolded. And the world rightly judged such evil conduct in the light of God's teachings. This was the equivalent of burning someone at the stake was forbidden murder, pure and simple.

But at the same time most of the world failed to consider that during this war, for it is war, our air strikes, of which Lieutenant Moaz was a knowing part, also burned to death on a daily basis large numbers of ISIS adherents, not to mention the "collateral damage" deaths of civilians, in the fiery blasts of our bullets, bombs, and rockets. People just haven't had their noses rubbed into the suffering of tens of thousands of humans currently estimated to have died this way. We don't have to experience up close and personal their deaths the way we did Moaz Kasasbeh's gruesome immolation.

Instead, on the nightly news we occasionally see grainy and soundless remote camera footage of puff-like blasts taking out a building or vehicle. What we don't see are pictures of the grizzly aftermaths of the charred and blackened human bodies which are the product of those blasts; people who were burnt to death with our approval.

But what God would have us recognize is that we are equally as responsible for our side's acts, just as ISIS's is for theirs. We have already demonstrated that God is prepared to use human violence to achieve God's purposes when human conduct has left God no other alternative. However, as a corollary God is also teaching us that we have to judge ourselves by the same standards we judge others, including ISIS. If ISIS' adherents have left the world no alternative other than their violent destruction, then so be it. Nonetheless, in doing so God wants us to understand that we have to take ownership of our own actions. We have to recognize them for what they truly are, the violent and fiery destruction of our fellow man, even if those men are acting in gruesomely evil and grossly misguided ways that leave us no other alternative.

In that light, God is teaching us that this sort of conduct must be our very last resort. It can only be used when there is nothing else that will advance God's quest to prevent negative change such as ISIS would inflict on humanity because it is up to us to take responsibility for being the actors needed to stop the ISISs of the world from undermining God's positive purposes for humanity even if we have to use violence to do so. It is up to us to learn how to use our free will to determine when there is no other choice. But, in doing so, we have to understand that God's lessons apply to all of humanity equally, not just to those we consider the bad guys.

God can and does teach us about all of this and so much more, but God can't live these lessons for us. We have to do that for ourselves in order to truly learn from such experiences. Like any good teacher, God understands that it is the real-world choices and challenges we have to face using God's lessons and guidance that force us to grow and mature; to move from being animals to becoming what God wants real humans to be.

Therefore it would seem that one of God's preferred roles is that of teacher, producing positive human growth in the ethical, moral, and spiritual sense as a result of our free will learning of God's lessons. It would also appear that another of God's preferences is to only actively engage in direct control or direction of the greater course of human events when God's teachings and subliminal urgings aren't enough or don't work. Then, when our inaction or negative conduct leave God no other choices besides engineering outcomes that still advance God's plan and purpose for humanity, God becomes proactive in terms of intervention. Unfortunately, when we put God in this position the end result, more often than not, entails very significant real-world pain, suffering, and distress for many of us; World War II comes immediately to mind as one of the prime examples of this.

Thus, the lessons God teaches are like seeds. Once planted, in company with our positive cooperation and partnership, they blossom

182

into robust structures that pollinate the rest of the world and all its societies in their turn. And those societies don't have to be of any given faith, or any faith at all. This is because God's lessons are not confined to any one group and are never finished, even when our individual encounters with God the teacher are concluded here on Earth. They continue to instruct and spread out from each student's life and conduct, even after that student has left this earthly existence.

God is challenging us to not just passively avoid sin, but instead to actively live in accordance with everything God is and has taught us. We can either pass or fail these God-taught courses. It is up to us. God won't do it for us. But with some luck and effort as well as God's grace, we can eventually graduate into a heavenly education that will take us to what God really needs us to become if we are to share eternity with each other and with God.

All of this is not to say that God does not regularly interact with each of us in ways that go far beyond God's role as teacher. God does. God is not in any way aloof from humanity, individually or collectively. While God is ultimately unknowable, God is always actively present to each and every one of us. But as part of that presence, God constantly maintains a watchful balance between not overwhelming our free will ability to choose our individual and collective courses of action, and directly influencing what happens in our lives and human history.

To use another analogy, God is like a parent standing behind a small child with a hand gently resting on the back of the tot's head. The parent steers the child in the direction the parent wants the child to go, almost unnoticed by the kid as the parent does so. The youngster totters intently forward toward its goal, missing the fact that its parent has kept it from wandering into a street full of oncoming traffic and away from other dangers. As the parent does so, the parent, in this case God, still wants the child to learn and grow in positive ways on its own. Parents work in this way because, having absorbed this guidance, children ultimately need to grow and mature to the point they will be able to avoid the traffic on their own. Thus, they will develop the capacity for a direct, knowing, and free will relationship with the parent teacher, God.

Chapter Twenty

GOD IS - YOUR VOTE

So at this point in our analysis of what we can know about God, it is reasonable to admit that we have no direct proof of God's existence. There are no pictures, videos, or electronic scans of God. There is no recording of God's voice and no credible testimony from a living witness who has had a direct conversation with God or heard God deliver a speech, let alone testimony from a group of people who were given audible directions or commands by God. And needless to say, no living human has actually seen God.

But that is not the same as saying we have no proof that in fact God is. Because we do have a large body of circumstantial evidence that points inexorably to God's existence and God's love for humanity.

Many will want to say, however, that circumstantial evidence is not enough. They need "hard" direct evidence that would stand up in a court of law, as well as the court of public opinion. But circumstantial proof is just that sort of evidence. It is relied on in court case after court case all over the globe and in just about every recognized legal system humanity has ever developed. It is one of the lynchpins of the British common law system found all across the English-speaking world including in the US. It is likewise accepted and used in those countries whose laws grow out of the Napoleonic Code. It's found in Sharia, the legal system experiencing resurgence where fundamentalist Islam has gained sway. It is regularly used by the world's emerging international courts such as the International Court of Justice and the European Court of Human Rights.

Some non-lawyers don't understand what circumstantial evidence is when compared to direct evidence, so it may help to look at what United States Federal Courts instruct their juries about such evidence today:

"Direct evidence" is the testimony of a person who asserts that he or she has actual knowledge of a fact, such as an eyewitness. "Circumstantial evidence" is proof of a chain of facts and circumstances that tend to prove or disprove a fact. There is no legal difference in the weight you may give to either direct or circumstantial evidence."[56]

In writing about circumstantial evidence the United States Supreme Court has held that, "Circumstances although inconclusive, if separately considered, may, by their number and joint operation, especially when corroborated by moral coincidences, be sufficient to constitute conclusive proof."[57] The Supreme Court went on in another case to say, "The reason for treating circumstantial and direct evidence alike is both clear and deep rooted: circumstantial evidence is not only sufficient, but may also be more certain, satisfying and persuasive than direct evidence."[58] Therefore, according to the US Supreme Court, "Circumstantial evidence means simply that [you] take one fact that has been seen, that is produced before you by evidence and from that fact [or facts] you reason to a conclusion."[59]

So with this understanding of circumstantial evidence let's review the chain of events and facts we've been considering in the prior chapters of this book that point to the fact that God actually is, and let's look at them in the way a jury would consider circumstantial evidence as to whether God exists or not.

We, of course, start this chain of events review the smallest instant after the Big Bang occurred. That explosion was, so far as we know, a one-time event that happened 13.8 billion years ago. It certainly has not been repeated in our universe since then. And its origin is a total mystery. The best scientists can say is that it was the product of what they euphemistically term a singularity, something coming into existence out of nothing. We just know that it happened and in happening produced everything needed to create our universe and us. But the why and how all of that was brought about is unknowable to us.

But looking at the first smallest part of a second following that out-of-nothing eruption, we can analyze the sequence of events that science and history tell us happened from there on and led to today.

1. This first universe-wide explosion we now speak of as the Big Bang (keeping in mind that at that point in time when we say universe wide, we are talking about a space smaller than your fist) was a precisely balanced blast of unbelievably powerful energy. Its force ejected, like an endless cloud of suddenly created microscopic shrapnel, all the subatomic particles needed to form our universe. They were what would ultimately become matter. And this was a universe that at that instant was inflating at hyper-speeds that put the speed of light to shame.

2. We say precisely balanced because at that same instant the force of gravity also winked into existence with a countervailing and again carefully set strength and pull that opposed the explosive outward push of the Big Bang's blast. This balance between outward pushing explosive force and inward pulling gravitation strength had to be just right to keep all the newly created subatomic particles from either dispersing into a fog of nothingness or alternately collapsing back in on themselves producing the oblivion of one miniature solid universal lump. Instead something had to cause the two strengths to be just right in terms of a balance that allowed all of what became matter to spread across the rapidly expanding universe in density groupings that produced the formation of all the galaxies of suns and planets that now surround us. So what was that something – random chance or intelligent planning?

3. The composition of this plasma of particulate matter produced by the Big Bang was also finely balanced. It was made up of both

matter and antimatter particles, and when those two interact they annihilate each other leaving only heat and radiation. Physics tells us that at the point of creation the two particle types could easily have been equally balanced in terms of their numbers or there could have been an imbalance in favor of antimatter being in the majority. Had either of those events occurred our universe as we know it would never have come into existence. There was, however, a sufficient positive imbalance in favor of matter over antimatter to permit the creation of our universe as we now see it. Was this just a lucky break?

4. The matter particles that survived likewise had to be finally tuned in terms of the infinitesimally small amounts of mass and electronic charges each particle carried in varying degrees. Without the correct combination of mass and electrical charges the subatomic particles that remained could never have combined into the various different atoms that now make up the stuff of our visible universe. And once again those electron and mass-per-particle properties could well have been different resulting in no hydrogen, carbon, oxygen, iron, etc. that we are so dependent on for our existence. Was this just another lucky spin of the cosmic roulette wheel or was it the product of intelligent forethought?

5. Initially those particles formed vast clouds of hydrogen and helium. Today these two elements make up ninety-nine percent of all the detectable matter in the universe. The universe's remaining one percent of visibly detectable matter was synthesized from their initial states as hydrogen and helium into all the rest of the elements, starting with the carbon we need and are made of. But that synthesis couldn't have happened absent a set of extraordinary circumstances at the level of the carbon atom.

Carbon is the sixth element that formed; however, this atom's formation in any quantity posed an incredibly tricky problem. Given carbon's atomic structure a road block existed that should have precluded the creation of sufficient amounts of this vital element needed to build the rest of the different elements, as well as carbon-based life-forms such as ourselves. Something had to jimmy the system at that exact point to permit enough carbon formation for our universe's needs and the needs of our carbon-based biology here on Earth, and it did in the form of the appearance of a unique isotope of helium that had the ability to interact with the beryllium atom. That isotope in conjunction with the precise setting of the strong nuclear force (described below) got around the problem. Without the appearance of that isotope and the just right setting of the strong nuclear force, the rest of creation would have come to a crashing halt. So did we just win another galactic-wide lottery?

6. Driving the creation of the different elements at the level of individual atoms are two forces physicists have named the strong nuclear force and the weak nuclear force. The strong nuclear force binds all the universe's subatomic particles together into the various atoms we and the rest of the universe are made of, and its strength had to be set at the precisely correct force value to make all this work. That is a value strong enough to hold each atom together and yet at the same time allow some of those atoms to break apart in the sort of synthesis needed to become the other atomic elements in the successive sequence of atoms that make up the periodic table of elements.

7. The strong nuclear force's companion, the weak nuclear force controls the actual process of synthesis without which each succeeding element, such as carbon, oxygen, iron, etc., could not come into existence. It, too, had to be precisely calibrated. The strength or force levels of both the strong and weak nuclear forces could easily have been so different from their "correct" settings, which came into existence at the beginning of creation, that the various forms of atoms as we know them would never have occurred, meaning no universe in the form we now inhabit. Do we put this down to a double win at the cosmic craps table or did someone or something intentionally set the strength dials at the precisely correct settings?

8. While all of this was occurring the force of gravity came back into play. Acting on the ninety-nine percent of the helium and hydrogen

not undergoing decay, gravity began pulling on those atoms. It aggregated them into giant clusters all across the universe. Over time each of those clusters resolved themselves into individual suns that in turn grouped themselves into galaxies. When each sun-size ball of hydrogen, reacting to the combined gravitational pulls of all its component atoms, reached a sufficient density, the resultant critical mass detonated thermonuclear chain reactions. It was as if each were a gigantic hydrogen bomb that encompassed the entire structure, and each burst into individual suns, lighting the entire universe. At that point their thermonuclear fires began an unquenchable burn that slowly consumed each star's massive, but finite, hydrogen fuel supplies. And how long it would take those nuclear fires to use up all that hydrogen leaving a dead star depended on the strength setting of the force of gravity in the universe.

Using our Sun as an example, if the force of gravity were double what it is, our Sun's lifetime would have shrunk from its current approximate ten-billion-year life expectancy to a mere 100 million years. And given the fact that it has taken 4.5 billion years to get our Earth and us to where we are today, a hundred million years wouldn't have been nearly long enough for our evolutionary climb. So we can see that the force of gravity had to once again be precisely set, and it was, when it could have been far different, leaving no us. So just how often could we win this game of cosmic chance without getting some assistance from somewhere, when any one break in the sequence of wins would have meant disaster for the entire universe and all the life that was to be contained in it?

9. And then there is the mystery of dark matter and dark energy. Science has now demonstrated that these two things make up about ninety-five percent of everything that's in our universe and are vital to its existence. Dark matter provides both the gravitational forces needed to keep our spinning galaxies from fragmenting and flying apart, while at the same time forming a sort of invisible skeleton or latticework upon which all the universe's visible matter constellations and galaxies are strung.

Its companion, dark energy is what propels the universe's apparently never-ending expansion, and it again had to be just right in terms of the amount of that energy present in the universe and its strength setting. If that amount or strength had been greater that they are they would have so dominated the matter and radiation we can detect that our universe's structure as we now see it would not have been able to form.

Yet these two vital components of the universe can only be inferred since they are not presently directly detectable.

10. Leaping some nine billion years down the universe's chain of history, to about 4.5 billion years ago when our Sun burst into thermonuclear flare and our own solar system was formed, we need to focus on the facts that produced our planet, this ball of rock and other elements that became the system's third planet out from the Sun.

The first of those facts is Earth's nearly perfectly circular orbit smack in what has been dubbed the Goldilocks zone. That zone is the distance from the Sun where its solar rays can produce the surface temperatures here on Earth needed in order to be biologically compatible on a year-round basis with the requirements for cellular life-forms' evolution, and where the liquid water supplies critical to that evolution could exist on the same year-round time frames. Other of our sister planets have elliptical, as opposed to circular, orbits. If Earth mimicked their orbits we would have wandered in and out of the Goldilocks zone, fatally interrupting life's evolutionary processes. Likewise if Earth had settled in the orbit of Venus, the second planet out from the Sun, things would have been too hot for life, or in that of Mars, the fourth of our system's planets, it would have been too cold. Another chance accident or were we put in just the right place intentionally?

11. The Earth itself turns out to contain the right mix of elements we need, starting with a massive water supply and an overall mass whose weight could generate a gravitational field strong enough to retain a breathable atmosphere once it developed. So did we just get lucky again?

191

12. Next, a major part of our planet's mass is made up of a gigantic molten iron core. That core, when spun up by the Earth's twenty-four hour period of daily axial rotation (that at the equator reaches a speed of about 1000 miles per hour), generates an electro-magnetic field called a magnetosphere. This resultant energy field surrounds and encompasses our entire planet. In doing so it acts as an invisible shield that deflects the searing blasts of the Sun's solar winds that constantly stream past us. Thus, its shielding effect protects all biological life here on Earth from being fried to a crisp, which is what would otherwise happen were those undeflected winds to strike us head-on. At the same time our planetary period of daily rotation allows for a relatively even distribution of the Sun's warmth over the entire planet on a daily basis. By comparison, a number of our sister planets lack such a large iron core and have axial rotation periods of a hundred days or more for each revolution. If our Earth had inherited either or both of those attributes life as we know it would never have been possible here. How did we get so lucky?

13. But it took all of this a billion years or so to come together. At that point our world was covered by primordial oceans surrounding a relatively compact land mass. Within those planet-girdling seas chemical reactions started occurring in which proteins began to chemically reproduce themselves. As they did so another onetime event occurred, when one and only one set of those proteins morphed into a living cell that managed to survive long enough to biologically reproduce itself. And it was that first living cell and its offspring whose endless evolutionary reproduction, based on the one single strand of DNA code of that very first living cell, that has created all of the biological life that now exists or has ever existed here on our Earth. But why only one DNA code that has remained fixed, unchanging for the 3.5 billion years since the appearance of that first cell? Why not hundreds, thousands, or even millions of different DNA codes rising out of the appearance of other cells at about the same time all across those oceans? Wouldn't the conditions necessary for cellular appearance have been

relatively equal across vast stretches of those primordial waters? So how come this didn't happen over and over again unless something intervened to keep that from occurring? And if so, who or what are we talking about?

14. And what about that very first parent cell? What could cause inanimate matter to morph into an incredibly complex cell that could at one and the same time produce porous, yet tough, cell walls and membranes, ingest nutrients through those walls, then turn them into food energy, and then expel the leftover waste products? Something that could additionally manufacture proteins needed for the cell's growth and the growth of a second identical daughter cell – a cell which it then gave live birth to through the process of cell division? Thereafter this same process of division would be repeated over and over again by all their resulting offspring. And how was that very first cell endowed with the prototype or template for every subsequent living cell for all the life that came after it right up to today and us?

What physically produced all this incredible change was a DNA chemical alphabet of characters that could and would be encoded into three character "words" which could then be combined in algorithm like structures. It was these algorithms which conveyed intelligible information and instructions to the previously inanimate matter that in turn became cells. And it was these algorithm strings that combined to form that first cell's genome or chemical blueprint. But what generated that alphabet and resulting genome?

It is suspected that this very first parent cell was similar to what today we know as the simplest life form alive, bacteria. The most studied of these is the *E.coli* bacteria which has a genome consisting of about four million characters programed in the exact sequence needed to produce *E.coli*.

So how did that DNA alphabet of chemical characters and that very first genome come into being if they were not the product of intentional design and thought? It defies logic to assume that a four-million-character chemical alphabet string similar to the *E.coli* genome could have just randomly happened to line up in just the precisely right million plus "word" order to be able to instruct and control previously inanimate matter such that it would suddenly form a functional living cell. Pure chance would seem to be a statistical impossibility. There had to be intelligence behind its design and creation.

15. As that first cell and its replicated offspring not only survived but reproduced themselves over and over again billions upon billions of times, change did occur, just not in their basic underlying DNA code. That change took the form of cell mutation that over the ensuing 3.5 billion years led to all the branches and varied species of earthly life, both animal and vegetable, in nearly endless profusion. In fact the result was so overwhelming that a process of culling had to occur to keep all that life in balance. This took the form of extinction events of which there have been at least five in the last half billion years. Interspaced between these events were also patterns of species die-offs. While disastrous in terms of the species that disappeared, they were all balanced enough to ensure the continued upward flow of evolution on a trajectory of positive change that led to us. Was all this again just the luck of the draw, or was something picking and choosing these outcomes?

16. As far as we humans are concerned, the most critical of these extinction events occurred 65 million years ago at the end of the Cretaceous Period when the Earth was apparently struck by a massive asteroid. The resulting impact blast generated the equivalent of nuclear winter lasting for years. This in turn produced a vast die-off of the Earth's vegetation, disrupting the then existing food chain, and leading to the extinction of the top of that chain, the dinosaurs, along with many other species. But it didn't end all life. Thus, for our purposes, this event was miraculously just right - not too big and not too small.

17. One of the species that did survive this holocaust was that of a small mammalian creature whose line had already made it through all the preceding extinction events and die-offs. Over the following tens of millions of years this line evolved and morphed into all the

mammalian creatures that would fill the evolutionary voids left by the demise of the dinosaur and reptilian lines. Had those predator lines of carnivores not been removed from the scene by that just right bombardment from space, our mammalian evolutionary ancestors would have had no hope of growing through the mechanisms of evolution into what became our direct primate forbearers. So once again was this all just another chance happening in this chain of just right events?

18. However, as the upward spiraling waves of plants, fungi, and animals continued their evolutionary climb into more and more complexity, something or someone focused on one part of the branch that became bipedal primates. This is our line, the great apes. Thus, starting about ten million years ago that line threw off in sequence, first gorillas, then hominids, and finally three million years ago chimpanzees. Gorillas and chimps are our near cousins, whose DNA genome codes only differ from ours by less than two percent, but in the millions of years since these, our cousins, emerged on the scene they have remained more or less static in terms of evolutionary change. By contrast, over the last three to four million years our hominid line has made a startling series of at least eighteen to twenty evolutionary leaps of which twelve or so have led directly to us. Thus we have to ask what, in terms of evolutionary time frames, drove this incredibly rapid progression of leaps to larger and larger hominid creatures with bigger and bigger brains when our gorilla and chimp cousins in effect stood still?

19. The last of these leaps has been the one that produced *Homo sapiens*, humanity. We appeared on the world stage two hundred thousand or so years ago. In that span of time we've gone from cave dwellers to the moon. In doing so humans have spread themselves across all of our planet's land masses, adapting to an incredible variety of climates and living conditions. This is something almost no other of the larger biological creatures has done without human aid. So what makes us think our successful adaptations were done without intentional assistance from something outside of our own species?

20. Throughout all this, the world we have been given to live on contains a finely calibrated set of physical challenges in the forms of other animal predators, disease, geophysical menaces, dangerous weather, climate extremes, and intellectual challenges that have forced us to grow, mentally and otherwise, in order to survive as a species. Until now the combinations of all these have not been so overwhelming that they threatened our existence. Nor have they been so benign that humanity could just coast through existence ignoring them. In short, they have been just right, in the Goldilocks sense, to push us forward in our mental, moral, and spiritual growth as individuals and as societies and civilizations. Why is that?

21. Perhaps the most important challenge humanity has been forced to overcome is the sheer diversity of humanity itself. We have spread all across the face of the Earth in groupings just isolated enough to produce different languages, cultures, and beliefs. As a consequence, instead of one more or less common language, culture, and belief system, we have many. The resultant challenging imperfections of life and human interactions have contributed mightily to driving our growth. Was that by accident or a product of higher intent?

22. Out of all this has come a wide range of philosophical and religious systems. One of the most successful of these at teaching the love of God and the love of neighbors as we love ourselves is Christianity. Arguably, under God's direction it has also been the most successful in attracting adherents on a worldwide basis. As we've seen, it is now, in aggregate, the world's largest religion, yet the creation and spread of Christianity was dependent on human violence, intolerance, and God's use of those forces to advance God's purpose and plan for humanity, especially including the absorption of the tenets of love, forgiveness, and tolerance.

The history of Jesus and his fate starkly highlight this argument. Despite preaching and practicing a message of love, forgiveness, healing, and salvation, Jesus went out of his way to ensure that the ruling elites of Israel and their Roman overlords would feel threatened by his activities and despise him. In fact, they were so threatened that they made his death the most brutal form of public execution available to them, crucifixion. It would seem this was as God would have it, because God needed Jesus' death to be so certain and public that his subsequent resurrection would have an almost incalculable impact on his disciples and followers. They went from being a frightened rabble, to that of such fearless advocates that they could withstand everything that was to follow, as they spread Jesus' "New Way" across the Roman world and beyond. Had Jesus not died such a horribly violent and public death, but lived out a long life of teaching and healing confined to his little Jewish world, he and his followers would have ended up just another in a line of prophets and their disciples with a minor following limited to that nation. Thus, God, acting through Jesus, had to go to extremely brutal lengths to engineer that death and what was to follow. It's hard to view all this as having just been another accident.

23. However, once crucified, Jesus left behind just that, a relatively small band of followers who formed a new sect within Judaism. This core group was almost exclusively Jewish. They were people who wished to remain a part of their Jewish world. They intended to await their Messiah's much anticipated second coming in that world, as they continued to worship in Jerusalem's Temple and proclaim Jesus' message to their fellow Jews. This, however, did not serve God's purpose or plan. God needed Jesus' message to be forced out of that insular existence and driven into the surrounding gentile nations where sufficient numbers of their pagan populaces were finally ready, though they didn't yet know it, for the teachings of Jesus that his disciples would proclaim.

To achieve this God had to next ensure that the Israelite leadership would seek to stamp out this New Way and its followers, even though Judaism had a long history of living with, or at least tolerating, competing sect-like factions within its people; these included the Pharisees, Sadducees, Zealots, and Essenes among others. A number of these were messianic in focus and espoused much of what Jesus taught. But

197

instead of simply ignoring Jesus' disciples or tolerating them as they did the other existing sects, the Israelite leaders thinking they were doing God's will (which they were – just not for the purposes they thought God intended) bent every effort to suppress them. Their attacks included forbidding the promulgation of Jesus' teachings, lashings of his principal disciples, the stoning deaths of the new sect's leaders, and other violent punishments. This achieved just what God intended, in that it forced what became Christianity out of that small and insular Jewish nation, into the surrounding sea of pagan countries, where there were many gentiles ready to receive these new teachings of love, humility, forgiveness, and salvation.

24. But Jerusalem's Temple priests and the rest of the Sanhedrin leadership were not content to simply drive Jesus' followers into exile; instead, they deputized delegations to hound these heretics in an effort to continue their attempt to stamp out this heresy. The most famous of these was the one that included Saul of Tarsus, known to us today as Paul. As an ultra-zealous Pharisee, Paul was hell bent on exterminating this new sect. His ardor was such that, to do so, he was willing to forsake his previous life of study and pious observation of Moses' laws in Jerusalem, to answer the Sanhedrin's call. He would pursue these heretics into the heart of the unclean pagan world, but in this effort something almost unbelievable happened to him outside of Damascus in what is now Syria. It was there that literally overnight he went from one of the staunchest persecutors of Jesus' following to becoming one of their leading advocates and organizers. In what is as close to direct evidence as we have, Paul, in his own words, testifies in his *Letter to the Galatians* (1:11-17) that it was God's own voice that drove him to the ground as it commanded him to proclaim Jesus as Messiah to the gentile world.[60]

Under its compulsion, that voice led Paul on from city to city, organizing Christian house church after house church, and what physically kept him on the move was the violence Paul faced in each of these cities and towns. In his own words (*Second Letter to the Corinthians*

11:25-33) these sufferings included lashings, beatings with rods, ston-ing, and the constant threat of death, imprisonment, and a host of other dangers that prevented him from lingering in any one spot, no matter how much he might long to stop and rest. This, of course, was as God intended it to be in order to ensure the spread of God's message of love, forgiveness, and salvation that was Christianity.

25. Over the next two hundred and forty years Christians and Christianity struggled forward to the point that they constituted an es-timated ten percent of the Roman Empire's population. Even so, they were still an oppressed and widely scattered minority whose very ex-istence was threatened at that time by extreme official persecution in many parts of the Roman world. In the span of only eighteen years, they then went from this beleaguered status to that of being the official religion of the Empire, a blink of an eye in terms of historical timelines.

This was precipitated by the actions of one man we know as the Emperor Constantine. As a young man in his twenties, in 306 C.E Constantine inherited from his father, the reigning co-August Con-stantius, a Roman Army and the potential to rule the western half of the Empire. Taking on those who would contest his right to rule, Con-stantine fought a series of epic battles that by 324 C.E. left him in undisputed command of the entire Roman Empire from the Atlantic all the way east to the borders of Persia.

Though Constantine was a pagan, he experienced a vision or vi-sions in which a supreme deity, he would later identify as God, pro-claimed to him that he was to rule the entire Roman world. In 312 C.E. just before one of the most critical of his fights, known as the Battle of the Milvian Bridge outside of the city of Rome itself, Constantine be-lieved he saw another vision, a cross in the sky with the Latin words *In Hoc Signo Vinces,* which translates In This Sign Conquer, inscribed be-neath it. With this inspiration, the young Constantine had a cross fash-ioned from a long lance with a cross bar lashed to it and the whole thing gilded in gold. This symbol then led his army to victory at the bridge.

Interestingly, up to that time the cross had not been utilized as a symbol for Christianity. But it became so thereafter when the totally victorious now Emperor Constantine made Christianity the Empire's official religion. In doing so Constantine became at least nominally a Christian himself. And with the power of his Empire behind it, Christianity quickly went from its ten percent persecuted minority status to that of the religion of the majority of Rome's world population. As such it was able to survive the fall of Rome a little more than a hundred years later: all because of what Constantine at least believed to have been the divine intervention of God leading him and the Empire to this outcome. The logical question then becomes would Christianity be here today in its present historically dominant position absent such divine intervention?

26. Fast forwarding to our recent past, let's now focus on World War II and Nazi Germany where its megalomaniac leader, Adolf Hitler, and his ruthlessly evil henchmen had set out to conquer the world under their banner of German National Socialism. As part of this, they intended to wipe out Judaism and its people, blaming them as the root cause of all of Germany's ills. They also intended to destroy Christianity with its weak-kneed advocacy of meekness, brotherly love, and forgiveness. In its place they intended to substitute worship of Hitler as the new messiah and National Socialism as the true Christianity. With this agenda they quickly crushed Europe's continental democracies as well as the traditional Christian church in Germany itself.

Despite God's silent urgings to stand up to Hitler and Nazi Germany, which were heard and heeded by only a small minority of Christian church and lay leaders, along with an equally small group of political figures, Hitler was well on his way to achieving these diabolical goals. As he advanced, the rest of the world including a pacifistic United States stood by in futile anguish, leaving only Great Britain and its associated Commonwealth of related countries and colonies in arms facing Hitler to Germany's west. To the Nazis' east sat a sleeping giant, the quiescent Union of Soviet Socialist Republics. It was controlled by an even more barbarically evil dictator, Joseph Stalin, who, distrusting the European democracies and to avoid a war with Germany, was willing to supply Nazi needs for raw materials, food and oil. Stalin did so even as he conducted his own campaigns against some of his neighboring nations and against all religions, especially Christianity within the USSR's borders. This, it would seem, left Hitler free to work his will on the rest of Europe.

In this nearly best of all worlds for the Germans, someone or something induced Hitler to commit the blunder of blunders. Without having knocked Britain out of the war, in 1941 Hitler invaded the Soviet Union, expecting another easy victory. Instead he got a four-year-long war against someone who was more than his match as a brutal dictator, Stalin, and his equally ruthless Communist cadres. This faceoff between totalitarian states ate up Germany's armies at the rate of millions of men a year.

At almost the same moment something or someone also induced Imperial Japan, a third dictatorship, that had its own ambitions to conquer Asia, to execute what turned out to be an equally foolhardy sneak attack on the US fleet anchored on a December Sunday in Pearl Harbor, Hawaii. This act enraged a previously pacifistic America, which had a far greater war-fighting capacity than Japan. It drove the US into the war on a no-holds-barred basis. Hitler, thinking he might need Japanese help against Russia, gratuitously declared war on the US in support of his nominal ally, Japan – an act that brought the Americans into the European conflict as well. And the rest, as they say is history. Germany and the USSR ground each other down on the eastern front as only two dictatorships could do. The might of the US swung the war in Western Europe and the Pacific against both Japan and Germany. Ultimately the dire threats the three dictatorships posed to democracy, Christianity, and Judaism, and religion in general, were vanquished.

What caused the missteps of Germany, Japan, and ultimately the USSR in this tragic drama can only be guessed at, but it is not unreasonable to see the hand of a loving God, who had previously demonstrated a caring regard for humanity at work. What else could have engineered all the interlocking missteps and blunders on the part of a pack of dictators who had previously been so surefooted?

27. However, a different but positive example presents itself in the Cold War aftermath of WWII. It is one in which it would seem God's silent voice was heeded by three world leaders. Their positive actions produced critical impacts on all of humanity and our world. The first two of these were John F. Kennedy and his Soviet counterpart Nikita Khrushchev. With both sides locked in a Cold War and armed to the teeth with enough nuclear weaponry to wipe out human civilization, the two-world dominating enemy camps came within an eye blink of a cataclysmic nuclear exchange in what became known as the 1962 Cuban Missile Crisis.

In response to America's stationing nuclear-tipped ballistic missiles on the northern Turkish border pointed at the USSR, the Soviets tried to clandestinely introduce their own missiles onto the island of Cuba, one of their Communist client states. The US discovered this effort and instituted a naval blockade of Cuba that was certain to lead to armed conflict and probably nuclear war if the USSR tried to force its ships transporting the missiles through that blockade.

Major portions of the military and political advisors surrounding both Khrushchev and Kennedy pressed their respective leaders to not back down in the face of the other side's threats, and both leaders were urged to show their opponents just how tough they could be. But both the American president and the Soviet first secretary elected to find a way for both sides to back away from the confrontation without losing too much face. They listened to calmer voices including the silent one in their heads.

The third man in this trilogy was Mikhail Gorbachev who, as general secretary of the Communist Party and the Soviet head of state, over a multi-year period in the late 1980s and early 1990s let the Soviet empire more or less peacefully collapse. This removed, at least for a period of time, the Russian threat to world peace and religions such as Christianity. Gorbachev did this at great cost to his own people as well as to his own personal standing and power in Russia. What induced him to do so? Again we must consider the possibility of God's silent voice.

28. There are any number of other facts and events that we could insert into this chain of circumstantial evidence, but from this sample you get the picture. In fact, it is probably a good bet that, if you tried, each of you could add your own individual instances and facts that would qualify. All you have to do is to look into your own lives and personal histories to see what you might find as both large and small instances when that small, silent voice, God's voice, has directed you in unexpected ways that truly guided or changed your lives; likewise, as part of this examination, it's likely you'll find instances where you ignored or resisted that voice to your detriment.

Thus, as we follow this string of predominately circumstantial evidence, the United States Supreme Court's pronouncement one hundred fifty years ago in *The Reindeer* case that "Circumstances although inconclusive, if separately considered, may, by their number and joint operation, especially when corroborated by moral coincidences, be sufficient to constitute conclusive proof," jumps to mind. Each of the just enumerated factual events, especially when put together, suggests the possibility that they had to have been planned, orchestrated, and executed by an other-worldly intelligence with a specific positive intent and outcome in mind. Perhaps any one of them could have happened by accident or chance, but for all of them to line up in just the right order and magnitude needed to leave us sitting here today goes beyond mere chance or luck of the draw.

Instead, in combination they suggest that we and our world are the products of a very unlikely, but nonetheless real, billions-of-years long chain of cause and effect. That chain demonstrates the distinct possibility, if not probability, that they were the sum result of intelligent design on a magnitude that can only be attributed to the idea that God is real, actively engaged in our world and lives, and focused on growing humanity out of our animal heritage toward a future perfection only God can see. When, for example, coupled with what it took to both create and then spread Judaism and its teachings, Buddhism, Christianity, and Islam, it would seem that one of God's focuses within this pattern of growth is to teach us individually and collectively the lessons Jesus and others have espoused: namely love of each other and God along with the attributes of forgiveness, morality, trust, and sacrifice that are the hallmarks of Jesus' ministry to humanity. They should also remind us that God is not our servant even though God loves us; instead, we are the servants of God set here to advance the plan God has been so carefully nurturing.

Many of us may be the modern equivalent of the doubting Apostle Thomas who said he would only believe in a resurrected Messiah if he could physically touch him and put his hands in the crucified Christ's living wounds.[61] But, if you approach the evidence we have just laid out with an open mind, an honest unbiased evaluation will have to at least lead you to acknowledge the distinct possibility that God exists, that God is.

Therefore, you the reader constitute a sort of endless jury who must decide for yourselves the question of whether or not you believe the case has been made that God is. Your vote, if yes, will take the form of attempting to live by and believing the lessons God has spent such endless amounts of time patiently teaching us. However, if despite all these circumstantial facts, your answer is no, then, on the off chance that you might be wrong, and God really is, you might still consider living by these same lessons based on love and your own self-interest – plus it is just the right thing to do.

Chapter Twenty-One

WHAT HAVE WE LEARNED

God has a plan! And not only a plan; but also a purpose for humanity as one of the central parts of that plan. God's plan for humanity began to unfold with God's creation of our universe and then, in terms of humankind, our planet, Earth. This was followed by God's initiation of biological life here on Earth and that life's billions of years-long evolutionary climb from one single living cell to the vast cornucopia of life that has now spread across our entire globe.

God has spent more than three billion years lovingly sculpting that prodigious mass of living things into an interlocking whole. This biologically variegated array of life simultaneously supports and challenges humanity as a result of God's intentional design. God has done all this in the same way master gardeners and husbandry men build their plant and animal creations through the mechanisms of both natural and artificial selection.

At the current apex of this earthly creation, God has installed humanity despite all our selfish arrogance and flaws. But, notwithstanding this fact, the truth is that we, like life itself, are still a work in progress as we both cooperate with and fight God's sculpting process, even as God seeks to grow us toward a true humanity.

The fundamental lesson we need to take from all this is that we humans cannot dictate to God, or demand what God is and must be in order for God to be worthy of acceptance as our deity. God is what God is. The truth is that while we may try to define God, we can only do so in human concepts, terms, and words that are in the end totally inadequate for that job.

For example, many want to insist that God has to be perfect, but that demand is based on our human characterization of perfection, when, in fact, God is other than perfect, as humans define that term. This does not imply that God is imperfect or even less than perfect. It just means that our definition of perfect does not even come close to meeting the challenge of defining or explaining God and God's complexities.

Instead, we have to accept God for the incredibly loving being manifested to us through God's creations and actions. These include the evolution of our universe and our Earth, as well as the miraculously rapid rise of humanity itself when compared to the evolutionary advances of the rest of the chain of life that we spring from.

So God is who God is. No amount of human willpower, longing, argument, or projection of our thoughts and desires onto the divine will change that; therefore, instead of trying to do so, we have to try to tease out how God functions in our real-world setting. Based on what we can deduce from this effort, we can begin to see the path God is directing us along while at the same time attempting to achieve some glimmers of an understanding of God.

In this attempt at understanding we need to recognize that no matter what their original inspiration, that is what initially inspired them, all religions are human constructs. They are all human efforts to reach for the divine whether that is considered to be an all controlling principal, essence or universal soul such as the *Atman* that according to Hindu teaching directs all the gods and every individual, a state of being such as Buddhism's *nirvana,* the God of Judaism and Christianity, or Islam's Allah. The divine in turn uses all of these and more to reach back to humanity in all its faiths.

By doing so, the divine seeks to teach all of humanity, no matter our culture or origin, to not only reach toward God, but also to reach out for each other and jointly mirror to the world the holiness that is God, the divine, by whatever name or concept we known it. It is this mirrored reflection that will lead humanity toward the ultimate perfection God is seeking for us and allow us to successfully face the challenges that can only be overcome by all of humanity working together, despite all our different beliefs and cultures.

The fact that all religions are made up of belief systems and structures erected by humans based on what they understand the divine has been telling them, means that we are going to have to deal with paradox and contradiction within and between all beliefs. It means we will be torn between faith and doubt, between certainty and uncertainty. Endless numbers of books have been written trying to deal with these problems. One small example is the question for Jews and Christians as to what best reflects God, the one presented in the Hebrew Bible (which is partly included in the Christian Old Testament) or the God of the Christian New Testament? Or is it both? Another is whether we are all totally dependent on and under the control of the divine, or whether we are beings with total free will who are the captains of our own destinies? Or is the answer somewhere in between?

Another of the fundamental lessons presented to us, and which becomes apparent from what we now understand, is that as an eternal being, who functions on the short-term time scales of billions of years, God is very patient and not in a hurry. As such God employs processes of incremental change to achieve positive advances in God's creations. This also implies that God is forward thinking and focused on the future as opposed to the past. This is only logical since we also now know that there never was an Adam and Eve Garden of Eden perfect past that God could be trying to return us to.

God has the power to do anything God wants to do but has chosen to act in a more restrained manner. God has refrained from using those infinite powers to instantly form us into what God wants us to become or from perfectly designing us.

Rather, what God has elected to do is to lovingly allow us to participate in doing for ourselves, as part of a partnership with God. One of the purposes of this partnership is to grow us out of our inherited evolutionary animal legacy of selfish self-focus, into the far more perfect humanity we have the capacity to become. As part of this effort, God gives every human the opportunity to be the most each of us can be and to experience everything that lies along the spectrum of life's journey.

By doing this we come to learn something of what God both is and is not. Learning what God is not is, in many ways, as important as learning what God is. God is not evil, nor is God the author of evil. Evil only exists because of human conduct that we have the capacity to recognize as being evil.

God is not indifferent, nor is God capricious. Most importantly, God is not our servant. God does not exist to fetch and carry at our whim, desire, or demand. God has no need to please us. When God pours out blessings and protection on us individually and collectively, those are acts of grace, pure gifts as opposed to anything we have earned or are entitled to expect and demand.

It is we, God's servants, who must ultimately answer to God, our creator. One of the principal ways we serve God is as God's students. In that capacity we are obligated to first learn and then internalize and live by the lessons God has spent well more than two hundred thousand years trying to teach us. At their core these lessons all focus on love: first loving God, followed by loving each other as God loves us. Out of this love grows the positive characteristics and relationships that bind us together and bind us to God. The practice of this love grows us toward the humanity-wide perfection God seems to be leading us to at some distant point in our collective future.

The learning servant process is, in turn, a function of our free will choices. It is not something God will use God's capacity for direct command to force on us. For whatever reason, God both wants, and seems to need, us to come to it out of a full experiential understanding and voluntary election to completely be that love with all it entails and demands of us.

Having foregone the option of open compulsion because it apparently robs us of our free will, which God seems to prize so highly, God instead speaks to us in a series of indirect ways. The first of which is God's silent voice. The vast majority of us hear that voice at a deep non-vocal and subconscious level. It percolates up out of our psyches as approving love, suggestive thought, or nagging conscience. But recognize it or not, we still hear it and must act or not act on it as a result of our own individual free will decision-making processes.

God also teaches us through the living examples of those of us who best comprehend and act on God's subconsciously delivered messages. In addition, God utilizes the truths embedded in the story lines God has fostered down through the ages in writings such as the Bible, Torah, Koran, and other religious and philosophical texts. It is unimportant to God whether those story lines are factually totally true or not, so long as we absorb the truths that are contained in the message. Our understanding of each story's message is what concerns God, and the most centrally important of those messages is that of the love that the Bible and God's other inspired writings consistently seek to teach us.

As a part of this teaching process God intentionally employs ambiguity and uncertainty. God deliberately leaves us in doubt as to what the answers are to a number of key questions. The first of these is whether God actually exists. The second is whether there is life after our biological deaths, and if there is, what does that life consist of.

Clearly God could give us direct, definitive answers to each of these questions if it served God's purposes to do so; however, God has refrained from doing this because knowing yea or nay answers to each of these issues would negatively impact our growth toward what God wants us to become.

If we had certainty, based on unambiguous factual proof that God existed, we would want to simply wait for divine directions as to what to do at critical junctures in our individual lives and human history. If to the contrary, we had no doubt that there is no God we needed to fear answering to, human conduct would be far more brutal and selfish over a far wider spectrum of humanity than it already is.

In the same way, if we knew for sure that there was nothing after our lives here on Earth, we would be far more prone to grab what we could in the here and now, without as much thought or concern for how our actions affected others or the eternal costs to ourselves. If we had absolute proof of an eternal other worldly existence after our biological deaths, we'd be in a far greater hurry to get ourselves and those same others into that heavenly realm.

All of this would inhibit the ethical, moral, and spiritual growth of humans, growth that God seems to prize. Instead of definitive answers, God has expended great efforts teaching us to act on these issues out of faith and faith's companion, trust. This ambiguity fosters God's plan for us since it forces us to grow.

This plan of God's also requires death as a result of biological bodily shortcomings and degenerative failure of all cellular life. Without the biologically driven imperative to reproduce in order to perpetuate each species in the face of individual biological death, God's decision to advance life through evolution would be a non-starter; it wouldn't work. It is the evolutionary advances achieved by each life-form over endless generations of reproductive replication that generates the evolutionary advances in all life that has ever existed. It's what has taken us from that initial single living cell that popped into existence 3.5 billion years ago to what life is today. If all the preceding generations hadn't died, where would we be today? Certainly, not where we are now.

Given the necessity of this fact, God has chosen to use the inevitability of death in very important ways to advance human intellectual, moral, and spiritual growth. It is our recognition of the certainty of our individual biological deaths that has shaped all of human civilization. We have to plan and prepare for the care of those we love but must leave behind. We create institutions as well as establish legal, moral, and philosophical systems to protect them. We build the physical infrastructures that will help ensure the perpetuation of all that is important to us when we are no longer here. Under God's tutelage we have come to care about history, about what our descendants will think about our actions and memory after our deaths. Will they love us or hate us; bless us or curse us?

And since death is often the product of the multitude of dangers found in our environment and the result of human conduct, God has steered us through the intellectual, moral, and organizational efforts needed to tame these threats and curb our animal Instincts toward violence. All of this makes us grow toward the levels of a more perfect humanity God seems to have in mind for us.

In the midst of this God has also taught us the importance of self-sacrifice in the face of danger and death. In fact, God has grown humanity to the point that many of us are prepared to risk our own well-being and even our lives to save others, be they loved ones or total strangers. We understand the risks of loss and death when we do so, and are prepared to pay that price if we must out of the love God has taught us. We are also prepared to defy the same threat of death to advance and protect institutions, ideas, and beliefs we cherish, be they intellectual, religious, social, or political. Under God's leadership and tutelage we have come to understand that there are many things more important than our individual lives. The lives and teachings of Moses, Buddha, Jesus of Nazareth, and Mohammed are prime representations of all this.

Another important lesson God has taught us through the example of these four men is how to change and even break the existing rules humans have laid down when those rules no longer advance God's growth of each generation toward the more nearly perfect humanity God has in store for us. Jesus' own flaunting of the rules of his earthly time when those rules hindered humanity's advance up the road God has laid out are emblematic, shining illustrations of how to do just that. From such exemplars we need to come to the realization that God is constantly seeking positive change within humanity, both individually and collectively. It is our responsibility to effectuate that change in partnership with God when it is needed.

211

To achieve these ends God searches out agents of change, both human and otherwise, and when God identifies those agents, be they events, individuals, or groups, God utilizes a multitude of means including silent suggestion, voluntary election, identified self-interest, compulsion and, when necessary, violent persuasion to induce the effort required to produce that change.

In the midst of all this we mustn't lose sight of the pragmatic side of God's nature. Our Deity works within the framework of the capacities of the human material available in any given generation, flawed as it may be. And this includes us. God tailors what God presents to each human cohort, its content and complexity, to the parameters of what each is capable of understanding and absorbing. We have to keep this fact in mind as we consider the stories, precepts, and lessons God used to teach our predecessor generations, ones that have then been handed down by them to us.

What God has made self-evident from the beginning of biological life and through humanity's entire history, is that God is not prepared to allow life or mankind's desire to perpetuate the status quo to stagnate God's required advances towards God's goals for us. We humans can, in turn, use our free will to either elect to partner with God in these endeavors of advancement or oppose God's efforts. It is our choice. But when we refuse to cooperate or actively seek to thwart God's purpose and plan, God will find the means to push us back on track even if that pushing inflicts incredible amounts of pain and suffering on many if not all of us.

From all this we need to also recognize that God's love for humanity, for us, is not necessarily manifested in just the ways we might want that love to show itself based on our own often self-centered focus; instead, God uses life to lovingly challenge each of us to be more than we ever thought or dreamed we could be. In doing this God pushes us in ways similar to those that really good parents use to encourage, challenge, and often drive their children, out of love for them because they truly care. And they keep right on pushing even when those children resent them for doing so or rebel.

But why is our loving God doing all this? Why has God created this plan and so patiently implemented it? What is God's purpose for our universe and all the biological life that we know exists here on Earth and that we must assume is seeded across the rest of the nearly endless universe surrounding us? We know that this physical universe and all the biological life it contains will ultimately cease to exist in what for us is the far, far distant future but for our eternal God is not much more than a few blinks of an eye. So why has God gone to all this trouble?

The answer proposed by this author is that our universe is an incubator for souls, human and otherwise. Souls, it would seem are non-corporal structures whose existence we have to take on faith. Something which will preserve the essence of our sentient selves for time frames far longer than the longest period our universe can hope to exist; in fact a time period outside of time.

God apparently has a use for our souls, us, in what we think of as an otherworldly existence we refer to as heaven. But to withstand the pressures of such an eternal or near eternal existence with each other in the presence of God, we have to leave our animal heritages behind us. We have to become more like God in terms of the attributes that will define each of us. To name just a few of them, these will include love, patience, empathy, fidelity, caring, moral courage, trust, and the willingness to make the right, but difficult, decisions that we may face in such an otherworldly life based on the exercise of our free will. All of these, especially trust, are necessary ingredients in the positive changes God is constantly fostering among we humans both in this life and the life to come.

Let us hope that in the end, as God's students, we are up to the task and justify God's faith in us. In the meantime, it is enough that we can seek to know even this smallest fraction of what God is – we can know that God Is.

About the Author

Alan Graham Greer is one of the nation's leading board-certified trial lawyers. A graduate of the U.S. Naval Academy at Annapolis, Maryland, he served six years in the U.S. Navy, including a year in Vietnam, before earning his Juris Doctor at the University of Florida College of Law. Alan is also the author of Choices & Challenges.

Active in a variety of civic and public affairs projects, Alan has worked with the homeless for more than twenty years most recently as Chairman of the Board of Camillus House, one of South Florida's largest provider of care to that population.

He is also Chairman Emeritus of the Board of Directors of Friends of WLRN Public Radio and Television and a member of the WLRN Chairman's Circle. He has served as Special Counsel for the Florida Wildlife Federation, handling water and environmental litigation on a pro bono basis. He is Past Chairman of the Dade County Council of Arts & Sciences and Founding President of the Bay Front Park Management Trust.

Alan served as a member of the Speaker's Task Force on Water Issues to recommend a state policy on water resources and a Trustee of the University of Florida Law Center Association as well as a past member of the ABA Standing Committee on Professionalism. He is also a Fellow of The American College of Trial Lawyers.

Author's Note

Dear Reader,

I hope you enjoyed reading *God Is* as much as I enjoyed writing it. Please do me a favor and write a review for me on amazon. The reviews are important, and your support is greatly appreciated. I can be reached at agreer@richmangreer.com or Facebook for further discussion.

Thank you,

Alan G. Greer

Other Books

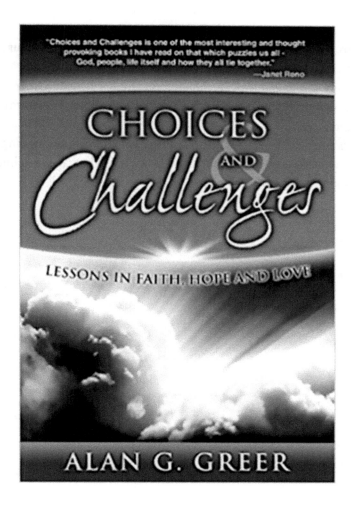

CHOICES

Challenges

LESSONS IN FAITH, HOPE AND LOVE

ALAN G. GREER

Choices & Challenges is a fresh way of looking at humanity's relationships with God and each other.

Based on the concept that while God loves us He does not exist to serve humanity but that we exist to serve God, Alan Greer confronts and debunks the atheistic notion that since our world and its multitude of religions are demonstrably imperfect with each and every one of us not being protected from all harm, God cannot exist because if He did He would have prevented all such imperfections.

Greer likewise challenges the other extreme that two to four thousand years ago God laid down a series of laws and rules that cannot under any circumstances be changed or broken. Instead Greer shows how God has taught us how and when to break those rules in favor of new and better ones.

Choices & Challenges demonstrates that God has an ongoing purpose for each of us and for humanity as a whole, that stretches into the far distant future.

Endnotes

[1] Based on the 2010 The World Almanac and http://www.100people.org/statistics_100stats.php?section-statistics

[2] "Return of the Tribes – The Resistance to Globalization Runs Deep" The Weekly Standard, September 4, 2006

[3] Long Walk To Freedom, Nelson Mandela, pg. 223, Little Brown and Company, 1994

[4] See, Thomas Aquinas, Quaestiones Desputatae, quoted in Joseph Pieper, The Silence of Saint Thomas Aquinas as translated by John Murray, S.J. and David O'Connor (Chicago: Henry Regnery 1965, pg. 69): Thomas Aquinas, Faith, Reason and Theology: Questions I-IV of his commentary on the De Trinitate of Bolthius as translated by Armand Mauer (Toronto) Pontifical Institute of Medieval Studies 1987 – Q 1, Article 2

[5] See Anthony de Millo, Awareness, Edited by Francis Stroud, S.J. Doubleday 1992

[6] We are using the word "being" here because it is the best English word we can come up with for something that exists, is sentient, and has conscious purpose on such infinite and eternal scales. But in the end God most certainly goes far beyond the concept of any mere being.

[7] At page 43. Davis was not, however, applying this equation to God. This author is.

[8] John 4:24

[9] Essential Cell Biology, Fourth Edition, pgs.4 – 5, Bruce Alberts and seven others, Garland Science, Taylor Francis Group, LLC, 2014.

[10] See id. Cell Biology, Fourth Edition generally.

[11] God's Undertaker, Has Science Buried God? Pg. 137, John C. Lennox, Lion Books, Lion Hudson plc, 2007.

[12] Richard Dawkins, The Greatest Show On Earth, pgs. 408-410, Free Press 2009

[13] A recent study led by Prof. Kunio Kaiho of Japan's Tohoku University suggests that the meteor plowed into a giant oil field under what is now the Yucatan Peninsula setting it ablaze. The resulting cloud of oil soot and ash would have encircled the globe and led to devastating climate change.

[14] Stephen Hawking and Leonard Mlodinow, The Grand Design, pg. 185, Bantam Books, 2010

[15] Recent discoveries in Russian Siberia suggest yet another hominid line in parallel with the Neanderthals which line is currently denominated Denisovans named after the cave in which the bone and tooth fragments which demonstrate its existence were found. Early indications suggest that based on the DNA extracted from these bone fragments the Denisovans contributed up to six percent of the DNA of modern Melanesians and Aboriginal Australians. Sapiens A Brief History of Humankind, Yuval Noah Harari, Harper Collins, 2015, pg. 16.

[16] Even more recently another potential human ancestor has been discovered in South Africa, Homo naledi that has many human characteristics as well as those of its ape ancestors. Its dates of existence and whether it was in or out of our evolutionary line are still open questions.

[17] Book of Amos 3:8

[18] Or as found in a more recent translation, "You seduced me, Lord, and I let myself be seduced, you were too strong for me and you prevailed...." The New American Bible, Revised Edition

Endnotes

(Jeremiah 20:7) Fireside Catholic Publishing, Wichita, KS and Devore & Sons, Inc. 2011

[19] Book of Jeremiah 20:7-9

[20] 1 Corinthians 9:16-17

[21] Genesis 7:11

[22] Pg. 70, The Evolution of God, Robert Wright, Little Brown and Company, 2009

[23] National Geographic 1999, www.nationalgeographic.com/flood.html

[24] Commentary on the Book of Genesis, pg. 1, The New American Bible, Catholic Bible Association of America (sponsored by the Bishops' Committee of the Confraternity of Christian Doctrine, Washington, D.C.) Thomas Nelson Publishers, 1970.

[25] An Interpretation of Religion, Human Responses to the Transcendent Second Edition 2004 Yale University Press

[26] The Hebrew Bible does not divide Samuel into two books as the Christian Bible does.

[27] Matthew 22:34-40/ Luke 10:25-28

[28] Luke 10:27, 30-37, Matthew 12:9-15; 15:1-20; 23:1-39

[29] Isaiah 43:1-14; Mark 16:15; Matthew 12:18-21; 13 in its entirety

[30] John 11:47-48

[31] Zechariah 9:9

[32] Luke 19:36-40

[33] John 6:15; Matthew 19:27-28

[34] Mark 15:26

[35] Kings 17:10-23, 2 Kings 2:1, 4:1-7, 4:42-44

[36] Pgs. 165-194, James Carroll, Constantine's Sword, The Church and the Jews, Houghton Mifflin Company, 2001.

[37] The Chi-Rho symbol is clearly depicted on contemporary coins and plaques honoring Constantine's subsequent confrontations with the Eastern Augustus Licinius only a few years after the Milvian Bridge battle.

[38] Id. at pgs. 189-92

[39] Id. at pgs. 192-93

[40] See generally William L. Shirer, The Nightmare Years, 1930-1940, Little Brown and Company 1984.

[41] Id. pgs. 151-155

[42] Id. pg. 154

[43] Peter Caddick-Adams, Snow and Steel, The Battle of the Bulge, 1944-45, pgs. 119 et seq., Oxford University Press 2015

[44] Pgs. 261-63 The Better Angles of Our Nature, Why Violence Has Declined, Steven Pinker, Viking, 2011

[45] March 2011 National Geographic Magazine Vol. 219, No.3, pg. 20

[46] See Id. pgs. 42 - 63

[47] Our Common Future, The World Commission On Environment and Development, Oxford University press, 1987

[48] Id. Footnote 46

[49] Pgs. 176-196 An Inconvenient Truth, Al Gore, Rodale 2006

[50] News article Miami Herald pg. A3, April 26, 2012, Journal Nature (April or May?) edition 2012

[51] Sept. 5, 2008 Scientific American article A deep Thaw, How much Will Vanishing Glaciers Rise Sea Levels, David Biello

[52] Id.

[53] Id.

[54] An Inconvenient Truth, pg. 73

[55] Id. pg. 81

[56] US Eleventh Circuit Court of Appeal approved Pattern Jury Instruction (2010) B4. This instruction epitomizes similar instructions given to juries all across the US Federal Court system throughout the United States.

[57] The Reindeer, 69 US 383, 401 (1864)

[58] Desert Palace Inc. v Costa, 539 US 90, 100 (2003)

[59] Hickory v. United States, 151 US 303, 310 (1894)

[60] In the Book of Acts 9:1-21 we have what would be considered its author, a third party's corroboration of Paul's testimony to being calling by God's voice and why he changed his position and conduct.

[61] John 20:24-29

God Is

a division of KCM Digital Media, LLC